Numerical PDE Analysis of the Blood Brain Barrier

Method of Lines in R

Numerical PDE Analysis of the Blood Brain Barrier

Method of Lines in R

William E Schiesser

Lehigh University, USA

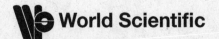
World Scientific

EW JERSEY · LONDON · SINGAPORE · BEIJING · SHANGHAI · HONG KONG · TAIPEI · CHENNAI · TOKYO

Published by

World Scientific Publishing Co. Pte. Ltd.
5 Toh Tuck Link, Singapore 596224
USA office: 27 Warren Street, Suite 401-402, Hackensack, NJ 07601
UK office: 57 Shelton Street, Covent Garden, London WC2H 9HE

Library of Congress Cataloging-in-Publication Data
Names: Schiesser, W. E., author.
Title: Numerical PDE analysis of the blood brain barrier : method of lines in R /
    by (author), William E Schiesser.
Description: New Jersey : World Scientific, 2018. | Includes bibliographical
    references and index.
Identifiers: LCCN 2018049400 | ISBN 9789813275799 (hardcover : alk. paper)
Subjects: | MESH: Blood-Brain Barrier | Mathematical Computing | Models, Theoretical
Classification: LCC QP375.5 | NLM WL 200 | DDC 573.8/621--dc23
LC record available at https://lccn.loc.gov/2018049400

British Library Cataloguing-in-Publication Data
A catalogue record for this book is available from the British Library.

For any available supplementary material, please visit
https://www.worldscientific.com/worldscibooks/10.1142/11144#t=suppl

Typeset by Stallion Press
Email: enquiries@stallionpress.com

Printed in Singapore

Contents

Preface

The remarkable functionality of the brain is made possible by the metabolism (chemical reaction) of oxygen (O_2) and nutrients in the brain. These chemical components for metabolism are supplied to the brain by an intricate blood circulatory system (vasculature). The rate of transfer from the blood to the brain tissue is determined by the blood brain barrier (BBB), which is the central topic of this book.

In particular, mathematical models are developed for mass transfer across the BBB based on partial differential equations (PDEs). The PDEs are derived from mass balances and computer routines in R are presented for the numerical (computer-based) solution of the PDEs.

The R routines and the associated numerical algorithms for computing the numerical solutions are discussed in detail. The discussion is introductory, without formal mathematics, e.g., theorems and proofs. The general methodology (algorithm) for numerical PDE solutions is the method of lines (MOL).

The tested and documented MOL R routines are available from a download link so the reader/researcher/analyst does not have to first study numerical algorithms and computer coding (programming), but rather, can proceed directly to the computation of PDE numerical solutions.

The models are based on each of the three sections, blood, brain and barrier, interconnected mathematically through boundary conditions. The computed concentration profiles of the reacting components are functions of time and space (location, coordinates) within the BBB system, that is, spatio-temporal solutions.

Numerical parameters (constants) in the PDEs can be varied to investigate the effect of mass transfer coefficients for the inter-section mass transfer rates, and diffusivities for the intra-section diffusion rates within the BBB sections. Also, the models can be extended from one component to multi components, including nonlinear effects. All of these variants are discussed in terms of detailed examples of BBB/PDE models.

The models are calibrated with parameter values that are estimated from the literature, but are considered realistic. The solutions for these parameter values, and the effect of variations in the parameters can be readily studied by executing the R routines on modest computers. The future reporting of research results should lead to refined values of the parameters.

Examples of computer-based analysis include variations of the mass transfer coefficients and diffusivities for movement of therapeutic drugs from the blood to the brain tissue. This type of study could, for example, be used to determine the effect of the drug molecule size which presently is a central research issue since the molecules are usually too large for efficient transfer through the BBB and the resulting low tissue concentration of these drugs renders them largely ineffective.

As a second example, the permeability of the barrier can be investigated. Permeability variations could result from brain trauma and age. An increase in the permeability could result in the brain tissue being susceptible to harmful substances and pathogens (bacteria, viruses) that are ordinarily excluded from the brain by the protection of the BBB.

In summary, the intent of the book is to provide computer-based models that can be used to study mass transfer through the BBB. This type of computer-based analysis might have application to the progression of neurodegeneration, e.g., Alzheimer's and Parkinson's disease, and the effectiveness of therapeutic drugs.

I hope some of these applications can be realized through the BBB/PDE models, and I would welcome comments about the application and usefulness of the models.

W. E. Schiesser
Bethlehem, PA 18015 USA
December 15, 2018

Chapter 1

PDE Model Formulation

(1.1) Introduction

The partial differential (PDE) models discussed in this book pertain to the transport across the blood brain barrier (BBB). These models include variations in space and time as depicted in Fig. 1.1 so they are *spatiotemporal* models.

The models have three sections: (1) the blood capillary (microvascular circulation), (2) the endothelial membrane, and (3) the brain tissue surrounding the capilliary. The endothelial membrane constitutes a barrier to the transfer of species from the blood to the brain, and thereby provides indispensable protection of the brain. In fact, when this barrier does not function adequately, perhaps because it has become permeable through trauma or age, the brain can no longer function normally, as manifest, for example, by the development of dementia, e.g., Alzheimer's or Parkinson's disease.

Each of the sections depicted in Fig. 1.1, (1), (2) and (3), is modeled by a PDE for the transport of one species, which, for example, might be O_2, a nutrient such as glucose, a neuropharmaceutical drug or a toxin or pathogen. Simultaneous transport of two or more species can be accommodated by using a set of PDEs for each species.

In the following discussion, the PDEs are derived in detail, and the associated parameters, initial conditions (ICs) and boundary conditions (BCs) are specified so that a complete

1

(well-posed) model is presented that is then programmed in R^1 as discussed in Chapter 2.

(1.2) Two PDE Model

The BBB system is depicted in Fig. 1.1.

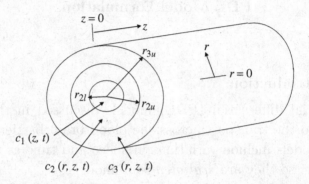

Figure 1.1: Schematic diagram of the BBB/PDE model variables

The PDE variables are defined in Table 1.1.

$c_1(z, t)$ blood concentration

$c_2(r, z, t)$ endothelial membrane concentration

$c_3(r, z, t)$ brain tissue concentration

---

[1]R is an open-source, quality scientific programming system, available from the Internet: http://cran.fhcrc.org/. Also, the ODE/DAE library deSolve [1] can be downloaded for use with R.

The editor Rstudio is recommended for use in working with R and for facilitating graphical output in established formats (e.g., eps, png, pdf): http://rstudio.org/. deSolve can be conveniently downloaded via Rstudio.

r	radial position
z	axial position
t	time

Table 1.1: BBB/PDE model dependent and independent
variables

Note that the concentration $c_1(z,t)$ is assumed to be uniform
across the capillary and therefore does not vary with r.

The following dimensions pertain to the system in Fig. 1.1:

r_{2l}, r_{2u}	lower, upper membrane radii
r_{3l}, r_{3u}	lower, upper tissue radii
z_l, z_u	lower, upper axial coordinates

Table 1.2: Dimensions of the BBB/PDE system

with $r_{2u} = r_{3l}$.

A two PDE model with dependent variables $c_1(a,t), c_2(r,z,t)$
is considered first as an introduction to spatiotemporal BBB
models. Then a three PDE model with dependent variables
$c_1(a,t), c_2(r,z,t), c_3(r,z,t)$ is considered in Chapter 3.

A mass balance on the blood with the incremental volume
$\pi r_1^2 \Delta z$ (with $r_1 = r_{2l}$, the radius of the capillary) gives

$$\pi r_1^2 \Delta z \frac{\partial c_1(z,t)}{\partial t} = \pi r_1^2 v_{c1}(z,t)c_1(z,t)|_z - \pi r_1^2 v_{c1}(z,t)c_1(z,t)|_{z+\Delta z}$$

$$-2\pi r_1 \Delta z(k_{12f}c_1(z,t) - k_{12r}c_2(r=r_1,z,t)) \qquad (1.1a)$$

The terms in eq. (1.1a) are

- $\pi r_1^2 \Delta z \dfrac{\partial c_1(z,t)}{\partial t}$: accumulation of the transported component in the incremental volume $\pi r_1^2 \Delta z$. If this term is neg-
ative (from the sum of the RHS terms), the concentration
of the component decreases (is depleted) with time.
- $\pi r_1^2 v_{c1}(z,t)c_1(z,t)|_z$: flow (convection) at velocity v_{c1}[2] of
the transported component into the incremental volume
at z.
- $-\pi r_1^2 v_{c1}(z,t)c_1(z,t)|_{z+\Delta z}$: flow (convection) at velocity
v_{c1} of the transported component out of the incremental
volume at $z + \Delta z$.
- $-2\pi r_1 \Delta z(k_{12f}c_1(z,t) - k_{12r}c_2(r = r_1, z, t))$: rate of mass
transfer of the transported component between the capil-
lary and the inner surface of the membrane. The transfer
area is $-2\pi r_1 \Delta z$ and the forward and reverse mass transfer coefficients are k_{12f}, k_{12r}, respectively (both positive).
If this term is negative, the transfer is from the blood
to the membrane, which reduces $\dfrac{\partial c_1(z,t)}{\partial t}$; the forward
transfer from the blood is greater than the reverse

---

[2]v_{c1} is assumed to be uniform (constant) across the capillary.
This is an idealization since the velocity will have a radial profile
with zero velocity at the capillary wall (the interface between
the blood and the membrane) and a maximum velocity at the
capillary centerline. If the blood is assumed to be Newtonian and
in laminar flow, the profile will be parabolic with the maximum
centerline velocity twice v_{c1}.

If a radial velocity profile is included, the blood concentration
will be $c_1(r, z, t)$ (rather than $c_1(z, t)$) and the r variation will
require radial gridding, which will add to the the MOL/ODE
system. By using v_{c1} rather than a radial velocity profile, the
overall MOL/ODE system is maintained at a more manage-
able size.

transfer to the membrane). This test of the sign of this term is essential in formulating the PDE. An incorrect sign will give a solution that physically is clearly nonsensible.

If eq. (1.1a) is divided by the incremental volume $\pi r_1^2 \Delta z$ (the coefficient of $\dfrac{\partial c_1(z,t)}{\partial t}$), after minor rearrangement,

$$\frac{\partial c_1(z,t)}{\partial t} = \frac{v_{c1}(z,t)c_1(z,t)|_{z+\Delta z} - v_{c1}(z,t)c_1(z,t)|_z}{\Delta z}$$
$$- \frac{2}{r_1}(k_{12f}c_1(z,t) - k_{12r}c_2(r=r_1,z,t)) \qquad (1.1\text{b})$$

For $\Delta z \to 0$, eq. (1.1b) becomes a PDE

$$\frac{\partial c_1(z,t)}{\partial t} = -\frac{\partial v_{c1}(z,t)c_1(z,t)}{\partial z}$$
$$- \frac{2}{r_1}(k_{12f}c_1(z,t) - k_{12r}c_2(r=r_1,z,t)) \qquad (1.1\text{c})$$

Eq. (1.1c) is the first PDE of the 2×2 system, that is, 2 PDEs in 2 unknowns, $c_1(z,t), c_2(r,z,t)$. Therefore a second PDE is required, which comes from a mass balance on an incremental volume, $2\pi r \Delta r \Delta z$[3].

$$2\pi r \Delta r \Delta z \frac{\partial c_2(r,z,t)}{\partial t} = 2\pi r \Delta z(-D_2(c_2,r,z,t)\frac{\partial c_2(r,z,t)}{\partial r})|_r$$
$$-2\pi r \Delta z(-D_2(c_2,r,z,t)\frac{\partial c_2(r,z,t)}{\partial r})|_{r+\Delta r} \qquad (1.1\text{d})$$

The terms in eq. (1.1d) are

- $2\pi r \Delta r \Delta z \dfrac{\partial c_2(r,z,t)}{\partial t}$: accumulation of the transported component in the incremental volume $2\pi r \Delta r \Delta z$. If this

---

[3]The incremental volume is $\pi(r+\Delta r)^2\Delta z - \pi r^2 \Delta z = \pi(r^2 + 2r\Delta r + \Delta r^2 - r^2)\Delta z \approx 2\pi r \Delta r \Delta z$ for small Δr.

term is negative (from the sum of the RHS terms), the concentration of the component decreases (is depleted) with time.

- $2\pi r \Delta z(-D_2(c_2, r, z, t)\dfrac{\partial c_2(r, z, t)}{\partial r})|_r$: diffusion into the incremental volume at r. This rate of diffusion is according to *Fick's first law*[4],

$$q_r = -D_2(c_2, r, z, t)\dfrac{\partial c_2(r, z, t)}{\partial r}$$

The minus is required so that the diffusion is in the direction of decreasing $c_2(r, z, t)$. That is, if $\dfrac{\partial c_2(r, z, t)}{\partial r} < 0$, $q_r > 0$.

- $-2\pi r \Delta z(-D_2(c_2, r, z, t)\dfrac{\partial c_2(r, z, t)}{\partial r})|_{r+\Delta r}$: diffusion out of the incremental volume at $r + \Delta r$.

If eq. (1.1d) is divided by the incremenatal volume $2\pi r \Delta r \Delta z$ (the coefficient of $\dfrac{\partial c_2(r, z, t)}{\partial t}$), after minor rearrangement,

$$\dfrac{\partial c_2(r, z, t)}{\partial t} =$$

$$\dfrac{1}{r \Delta r}\left((r + \Delta r)D_2(c_2, r, z, t)\dfrac{\partial c_2(r, z, t)}{\partial r}\Big|_{r+\Delta r}\right.$$

$$\left. -rD_2(c_2, r, z, t)\dfrac{\partial c_2(r, z, t)}{\partial r}\Big|_r\right) \tag{1.1e}$$

For $\Delta z \to 0$, eq. (1.1e) becomes a PDE

$$\dfrac{\partial c_2(r, z, t)}{\partial t} = \dfrac{1}{r}\dfrac{\partial\left(rD_2(c_2, r, z, t)\dfrac{\partial c_2(r, z, t)}{\partial r}\right)}{\partial r}$$

---

[4]Other forms of q_r can be considered at this point that could include alternative transport mechanisms.

and with $D_2(c_2(r, z, t), r, z, t) = D_2(c_2(r, z, t))$

$$\frac{\partial c_2(r, z, t)}{\partial t} = \frac{1}{r} \frac{\partial \left(r D_2(c_2(r, z, t)) \frac{\partial c_2(r, z, t)}{\partial r} \right)}{\partial r} \qquad (1.1f)$$

If the product term in r is differentiated,

$$\frac{\partial c_2(r, z, t)}{\partial t} = D_2(c_2(r, z, t)) \left(\frac{\partial^2 c_2(r, z, t)}{\partial r^2} + \frac{1}{r} \frac{\partial c_2(r, z, t)}{\partial r} \right)$$
$$+ \frac{\partial D_2(c_2(r, z, t))}{\partial c_2(r, z, t)} \left(\frac{\partial c_2(r, z, t)}{\partial r} \right)^2 \qquad (1.1g)$$

For the special case $D_2(c_2(r, z, t)) = D_2$ (a constant), $\frac{\partial D_2(c_2(r, z, t))}{\partial c_2(r, z, t)} = 0$, and eq. (1.1g) is

$$\frac{\partial c_2(r, z, t)}{\partial t} = D_2 \left(\frac{\partial^2 c_2(r, z, t)}{\partial r^2} + \frac{1}{r} \frac{\partial c_2(r, z, t)}{\partial r} \right) \qquad (1.1h)$$

Eq. (1.1h) is *Fick's second law* in 1D cylindrical coordinates (with the angular and axial components neglected).

Eqs. (1.1c) and (1.1h) are first order in t and each requires one IC.

$$c_1(z, t = 0) = c_{10}(z); \quad c_2(r, z, t = 0) = c_{20}(r, z) \qquad (1.2a,b)$$

Eq. (1.1c) is first order in z and requires one BC.

$$c_1(z = 0, t) = c_{1e}(t) \qquad (1.3)$$

Eq. (1.1h) is second order in r and requires two BCs.

$$D_2 \frac{c_2(r = r_{2l}, z, t)}{\partial r} = -(k_{12f} c_1(z, t) - k_{12r} c_2(r = r_{2l}, z, t)) \qquad (1.4a)$$

$$D_2 \frac{c_2(r = r_{2u}, z, t)}{\partial r} = 0 \qquad (1.4b)$$

Eq. (1.4a) states that the rate of diffusion at the boundaries of the epithelial membrane (LHSs) equals the rate of mass transfer stated in terms of the mass transfer coefficients k_{12f}, k_{12r}. Again, a check on the signs of the LHS and RHS terms is essential. For example, for net transfer from the blood to the lower boundary of the membrane, the terms in eq. (1.4a) are negative. Eq. (1.4b) is a no flux (homogeneous Neumann) BC at the upper boundary of the membrane.

Eqs. (1.1c,h) to (1.4) constitute a 2×2 PDE system programmed in Chapter 2.

Reference

[1] Soetaert, K., J. Cash, and F. Mazzia (2012), *Solving Differential Equations in R*, Springer-Verlag, Heidelberg, Germany.

Chapter 2

PDE Model Computer Implementation

(2.1) Introduction

The computer implementation of the PDE models developed in Chapter 1 is considered in this chapter. The coding (programming) of a series of routines in R is explained and the output from the routines, both numerical and graphical, is discussed.

(2.2) Two PDE Model

A main program for the two PDE model of eqs. (1.1c,h) to (1.4) is considered first.

(2.2.1) Main program

A main program for eqs. (1.1c,h) to (1.4) follows.

```
#
# Two PDE BBB model
#
# Delete previous workspaces
  rm(list=ls(all=TRUE))
#
# Access ODE integrator
  library("deSolve");
#
# Access functions for numerical solution
  setwd("f:/BBB/chap2/ex1");
```

9

```
  source("pde1a.R");
  source("dss004.R");
  source("dss044.R");
  source("vanl.R");
#
# Model parameters
     u10=0;    u20=0;
  D2=1.0e-06;  v_c1=1;
     c1e=1;
     k12f=1;  k12r=1;
#    k12f=0;  k12r=0;
#
# Initial condition
  nz=21;nr=6;
  nznr=nz*nr;
  u0=rep(0,nz+nznr);
  for(iz in 1:nz){
    u0[iz]=u10;
  for(ir in 1:nr){
    izir=(iz-1)*nr+ir;
    u0[nz+izir]=u20;
  }
  }
#
# Grid in z
  zl=0;zu=1;
  z=seq(from=zl,to=zu,by=(zu-zl)/(nz-1));
#
# Grid in r for c2
  r2l=1.0e-03;r2u=2.0e-03;
  r2=seq(from=r2l,to=r2u,by=(r2u-r2l)/(nr-1));
#
# Interval in t
  t0=0;tf=2;nout=6;
```

```
# t0=0;tf=1;nout=6;
  tout=seq(from=t0,to=tf,by=(tf-t0)/(nout-1));
  ncall=0;
#
# ODE integration
  out=lsodes(y=u0,times=tout,func=pde1a,
      sparsetype="sparseint",rtol=1e-6,
      atol=1e-6,maxord=5);
  nrow(out)
  ncol(out)
#
# Store solution
  c1 =matrix(0,nrow=nz,ncol=nout);
  c2l=matrix(0,nrow=nz,ncol=nout);
  c2u=matrix(0,nrow=nz,ncol=nout);
  t=rep(0,nout);
  for(it in 1:nout){
  for(iz in 1:nz){
     c1[iz,it]=out[it,iz+1];
    c2l[iz,it]=out[it,nz+1+(iz-1)*nr+1];
    c2u[iz,it]=out[it,nz+1+(iz-1)*nr+nr];
        t[it]=out[it,1];
  }
  c1[1,it]=c1e;
  }
#
# Display ncall
  cat(sprintf("\n\n ncall = %2d",ncall));
#
# Display numerical solution
  for(it in 1:nout){
  cat(sprintf(
    "\n\n     t      z    c1(z,t)    c2l(r,z,t)
     c2u(r,z,t)"));
```

```
  izv=seq(from=1,to=nz,by=2);
  for(iz in izv){
    cat(sprintf(
      "\n%7.2f%7.2f%10.4f%12.4f%12.4f",
      t[it],z[iz],c1[iz,it],c2l[iz,it],
      c2u[iz,it]));
  }
  }
#
# Plot numerical solutions
#
# c1(z,t)
  par(mfrow=c(1,1));
  matplot(
    x=z,y=c1,type="l",xlab="z",
    ylab="c1(z,t)",xlim=c(zl,zu),
    lty=1,main="",lwd=2,col="black");
#
# c2(r=r2l,z,t)
  par(mfrow=c(1,1));
  matplot(
    x=z,y=c2l,type="l",xlab="z",
    ylab="c2(r=r2l,z,t)",xlim=c(zl,zu),
    lty=1,main="",lwd=2,col="black");
#
# c2(r=r2u,z,t)
  par(mfrow=c(1,1));
  matplot(
    x=z,y=c2u,type="l",xlab="z",
    ylab="c2(r=r2u,z,t)",xlim=c(zl,zu),
    lty=1,main="",lwd=2,
    col="black");
```

Listing 2.1: Main program for eqs. (1.1c,h) to (1.4)

We can note the following details about Listing 2.1.

- Previous workspaces are deleted.

```
#
# Two PDE BBB model
#
# Delete previous workspaces
  rm(list=ls(all=TRUE))
```

- The R ODE integrator library deSolve is accessed. Then the directory with the files for the solution of eqs. (1.1c,h) to (1.4) is designated. Note that setwd (set working directory) uses / rather than the usual \.

```
#
# Access ODE integrator
  library("deSolve");
#
# Access functions for numerical solution
  setwd("f:/BBB/chap2/ex1");
  source("pde1a.R");
  source("dss004.R");
  source("dss044.R");
  source("van1.R");
```

pde1a.R is the routine for the method of lines (MOL) approximation[1] of PDEs (1.1c,h) (discussed

---

[1]The method of lines is a general numerical procedure for the integration of PDEs in which the spatial (boundary value) derivatives are approximated algebraically, e.g., by finite differences (FD), finite elements, weighted residuals, least squares, spectral methods, flux limiters (FL).

Then an initial value variable remains, usually time, and with one independent variable, the PDEs are approximated by a system of initial value ODEs that are integrated numerically, usually with a library ODE integrator. The present MOL

subsequently). dss004, dss044 (Differentiation in Space Subroutine) are library routines for calculating first and second derivatives in r by finite differences (FDs). van1 is a library routine for calculating first derivatives in z. The coding and use of these routines is discussed subsequently.

- The parameters of eqs. (1.1c,h) to (1.4) are defined numerically.

```
#
# Model parameters
      u10=0;    u20=0;
  D2=1.0e-06;  v_c1=1;
      c1e=1;
      k12f=1;   k12r=1;
#     k12f=0;   k12r=0;
```

A second case with $k_{12f} = k_{12r} = 0$ is considered subsequently.

- ICs (1.2a,b) are defined.

```
#
# Initial condition
  nz=21;nr=6;
  nznr=nz*nr;
  u0=rep(0,nz+nznr);
  for(iz in 1:nz){
    u0[iz]=u10;
  for(ir in 1:nr){
    izir=(iz-1)*nr+ir;
    u0[nz+izir]=u20;
  }
  }
```

---

analysis is based on FD and FL approximations of the spatial derivatives.

IC (1.2a) is placed in u0[1] to u0[nz] as u10. IC (1.2b) is placed in u0[nz+1] to u0[nz+nz*nr] as u20.

- A spatial grid of nz=21 points is defined for $z_l = 0 \leq z \leq z_u = 1$, so that z=0,0.05,...,1.

```
#
# Grid in z
  zl=0;zu=1; .
  z=seq(from=zl,to=zu,by=(zu-zl)/(nz-1));
```

- A spatial grid of nr=6 points is defined for $r_{2l} = 1 \times 10^{-3} \leq r_2 \leq r_{2u} = 2 \times 10^{-3}$, so that $r_2 = 0.001, 0.0012, \ldots, 0.002$.

```
#
# Grid in r for c2
  r2l=1.0e-03;r2u=2.0e-03;
  r2=seq(from=r2l,to=r2u,by=(r2u-r2l)/(nr-1));
```

The small number of points in r2, nr=6, was selected with the total number of ODEs in mind, $21 + (21)(6) = 147$. That is, the number of ODEs increases rapidly with the product $(nz)(nr)$. In this case, for each z, 1 MOL/ODE is specified for $c_1(z,t)$ and 6 MOL/ODEs are specified for $c_2(r,z,t)$ for a total of $(21)(7) = 147$ MOL/ODEs.

- An interval in t of 6 points is defined for $0 \leq t \leq 2$ so that tout=0,0.4,...,2.

```
#
# Interval in t
  t0=0;tf=2;nout=6;
# t0=0;tf=1;nout=6;
  tout=seq(from=t0,to=tf,by=(tf-t0)/(nout-1));
  ncall=0;
```

A second case with `tf=1` is considered subsequently. The counter for the calls to the ODE/MOL routine `pde1a` is also initialized.

- The system of 147 MOL/ODEs is integrated by the library integrator `lsodes`[2] (available in `deSolve`). As expected, the inputs to `lsodes` are the ODE function, `pde1a`, the IC vector `u0`, and the vector of output values of t, `tout`. The length of `u0` (e.g., 147) informs `lsodes` how many ODEs are to be integrated. `func,y,times` are reserved names.

```
#
# ODE integration
  out=lsodes(y=u0,times=tout,func=pde1a,
    sparsetype="sparseint",rtol=1e-6,
    atol=1e-6,maxord=5);
  nrow(out)
  ncol(out)
```

The numerical solution to the ODEs is returned in matrix `out`. In this case, `out` has the dimensions $nout \times nz + (nz)(nr) + 1 = 6 \times 21 + (21)(6) + 1 = 148$, which are confirmed by the output from `nrow(out),ncol(out)` (included in the numerical output considered subsequently).

The offset $147+1 = 148$ is required since the first element of each column has the output t (also in `tout`), and the

---

[2]`lsodes` (Livermore Solver for Ordinary Differential Equations Sparse) is an initial value ODE integrator in `deSolve` that is based on sparse matrix processing of the ODE Jacobian matrix. Although `lsodes` requires added calculations internally for the sparse matrix algorithm, the additional calculations usually result in an efficient numerical solution of the ODE system, particularly as the size and stiffness of the ODE system increase.

$2, \ldots, 147 + 1 = 2, \ldots, 148$ column elements have the 147 ODE solutions.

- Selected solutions of the 147 ODEs returned in out by lsodes are placed in arrays c1, c21, c2u.

```
#
# Store solution
  c1 =matrix(0,nrow=nz,ncol=nout);
  c21=matrix(0,nrow=nz,ncol=nout);
  c2u=matrix(0,nrow=nz,ncol=nout);
  t=rep(0,nout);
  for(it in 1:nout){
  for(iz in 1:nz){
    c1[iz,it]=out[it,iz+1];
    c21[iz,it]=out[it,nz+1+(iz-1)*nr+1];
    c2u[iz,it]=out[it,nz+1+(iz-1)*nr+nr];
        t[it]=out[it,1];
  }
  c1[1,it]=c1e;
  }
```

Again, the offset iz+1 is required since the first element of each column of out has the value of t. $c_1(z = 0, t) = $ c1[1,it]=c1e*(1-exp(-t_c1*t[it])) sets BC (1.3) since this value is not returned by lsodes (it is defined algebraically in pde1a and not by the integration of an ODE at $z = 0$).

$c_2(r = r_{2l}, z, t), c_2(r = r_{2u}, z, t)$ are placed in c21, c2u.

- The number of calls to pde1a is displayed at the end of the solution.

```
#
# Display ncall
  cat(sprintf("\n\n ncall = %2d",ncall));
```

- $t, z, c_1(z, t), c_2(r = r_{2l}, z, t), c_2(r = r_{2u}, z, t)$ are displayed for every second value of z from by=2.

```
#
# Display numerical solution
  for(it in 1:nout){
  cat(sprintf(
    "\n\n      t      z   c1(z,t)    c21(r,z,t)
     c2u(r,z,t)"));
  izv=seq(from=1,to=nz,by=2);
  for(iz in izv){
    cat(sprintf(
      "\n%7.2f%7.2f%10.4f%12.4f%12.4f",
      t[it],z[iz],c1[iz,it],c21[iz,it],
      c2u[iz,it]));
  }
}
```

- $c_1(z, t), c_2(r = r_{2l}, z, t), c_2(r = r_{2u}, z, t)$ are plotted as a function of z with t as a parameter.

```
#
# Plot numerical solutions
#
# c1(z,t)
  par(mfrow=c(1,1));
  matplot(
    x=z,y=c1,type="l",xlab="z",
    ylab="c1(z,t)",xlim=c(zl,zu),
    lty=1,main="",lwd=2,col="black");
#
# c2(r=r21,z,t)
  par(mfrow=c(1,1));
  matplot(
    x=z,y=c21,type="l",xlab="z",
```

```
        ylab="c2(r=r2l,z,t)",xlim=c(zl,zu),
        lty=1,main="",lwd=2,col="black");
  #
  # c2(r=r2u,z,t)
    par(mfrow=c(1,1));
    matplot(
      x=z,y=c2u,type="l",xlab="z",
      ylab="c2(r=r2u,z,t)",xlim=c(zl,zu),
      lty=1,main="",lwd=2,
      col="black");
```

This completes the main program. The ODE/MOL routine
pde1a called by lsodes is considered next.

(2.2.2) ODE/MOL routine

pde1a follows.

```
  pde1a=function(t,u,parms){
#
# Function pde1a computes the t derivatives
# c1t(z,t), c2t(r,z,t)
#
# One vector to one vector, one matrix
  c1=rep(0,nz);
  c2=matrix(0,nrow=nz,ncol=nr);
  for(iz in 1:nz){
    c1[iz]=u[iz];
  for(ir in 1:nr){
    izir=(iz-1)*nr+ir;
    c2[iz,ir]=u[izir+nz];
  }
  }
#
# c1 BC
  c1[1]=c1e;
```

```
#
# c1z
  c1z=vanl(zl,zu,nz,c1,v_c1);
#
# c2r
  c2r=matrix(0,nrow=nz,ncol=nr);
  for(iz in 1:nz){
    c2r[iz,]=dss004(r2l,r2u,nr,c2[iz,]);
  }
#
# c2 BCs
  for(iz in 1:nz){
    c2r[iz,1] =-(1/D2)*(k12f*c1[iz]-
                k12r*c2[iz,1]);
    c2r[iz,nr]=0;
  }
#
# c2rr
  nl=2;nu=2;
  c2rr=matrix(0,nrow=nz,ncol=nr);
  for(iz in 1:nz){
    c2rr[iz,]=dss044(r2l,r2u,nr,c2[iz,],
                     c2r[iz,],nl,nu);
  }
#
# c1t, c2t
  c1t=rep(0,nz);
  c2t=matrix(0,nrow=nz,ncol=nr);
  for(iz in 1:nz){
    c1t[iz]=-v_c1*c1z[iz]-(2/r2l)*
            (k12f*c1[iz]-k12r*c2[iz,1]);
  for(ir in 1:nr){
    c2t[iz,ir]=D2*(c2rr[iz,ir]+(1/r2[ir])*
               c2r[iz,ir]);
```

```
    }
    }
  c1t[1]=0;
#
# One vector, one matrix to one vector
  ut=rep(0,(nz+nznr));
  for(iz in 1:nz){
    ut[iz]=c1t[iz];
  for(ir in 1:nr){
    izir=(iz-1)*nr+ir;
    ut[nz+izir]=c2t[iz,ir];
  }
  }
#
# Increment calls to pde1a
  ncall<<-ncall+1;
#
# Return derivative vector
  return(list(c(ut)));
}
```

Listing 2.2: ODE/MOL routine for eqs. (1.1c,h) to (1.4)

We can note the following details about Listing 2.2.

- The function is defined.

    ```
      pde1a=function(t,u,parms){
    #
    # Function pde1a computes the t derivative
    # vector of c1t(z,t), c2t(r,z,t)
    ```

    t is the current value of t in eqs. (1.1c,h). u is the 147-vector of ODE/MOL dependent variables. **parm** is an argument to pass parameters to **pde1a** (unused, but required in the argument list). The arguments must be listed in the order stated to properly interface with 1sodes called in the main program of Listing 1.1. The derivative vector of the LHS of eqs. (1.1c,h) is calculated next and returned to. 1sodes.

- u is placed in one vector, c1, and one matrix, c2, to facilitate the programming of eqs. (1.1c,h) to (1.4).

```
#
# One vector to one vector, one matrix
  c1=rep(0,nz);
  c2=matrix(0,nrow=nz,ncol=nr);
  for(iz in 1:nz){
    c1[iz]=u[iz];
  for(ir in 1:nr){
    izir=(iz-1)*nr+ir;
    c2[iz,ir]=u[izir+nz];
  }
  }
```

The dimensions for the vectors and matrix are u(147), c1(21), c2(21,6).
- BC (1.3) is implemented.

```
#
# c1 BC
  c1[1]=c1e;
```

- $\dfrac{\partial c_1(z,t)}{\partial z}$ in eq. (1.1c) is computed by van1.

```
#
# c1z
  c1z=van1(zl,zu,nz,c1,v_c1);
```

c1z does not have to be allocated (with rep) since this is done by van1.

`vanl` is an implementation of the van Leer flux limiter that is designed specifically to approximate first order, convective derivatives, e.g., $\dfrac{\partial c_1(z,t)}{\partial z}$ in eq. (1.1c). Additional details about the van Leer and several other flux limiters are given in [2].

- The first derivative $\dfrac{\partial c_2(r,z,t)}{\partial r}$ in eq. (1.1h) is computed by `dss004`.

```
#
# c2r
  c2r=matrix(0,nrow=nz,ncol=nr);
  for(iz in 1:nz){
    c2r[iz,]=dss004(r2l,r2u,nr,c2[iz,]);
  }
```

The values of z are selected with the `for`. The 6 values of r are selected with the `,` subscript. Therefore, `c2r` is a 6-vector at a particular z.

- BCs (1.4a,b) are implemented (the subscripts `1,nr` correspond to `r2l,r2u`) at each z.

```
#
# c2 BCs
  for(iz in 1:nz){
    c2r[iz,1] =-(1/D2)*(k12f*c1[iz]-
                k12r*c2[iz,1]);
    c2r[iz,nr]=0;
  }
```

- The second derivative $\dfrac{\partial^2 c_2(r,z,t)}{\partial r^2}$ in eq. (1.1h) is computed by `dss044`.

```
#
# c2rr
  nl=2;nu=2;
```

```
  c2rr=matrix(0,nrow=nz,ncol=nr);
  for(iz in 1:nz){
    c2rr[iz,]=dss044(r2l,r2u,nr,c2[iz,],
                     c2r[iz,],nl,nu);
  }
```

Since BCs (1.4a,b) are Neumann, `nl=2;nu=2` is used before calling `dss044`. Additional details about `dss044` are available in Appendix A2.

- PDEs (1.4c,h) are programmed, and the derivatives in t are placed in `c1t, c2t`.

```
#
# c1t, c2t
  c1t=rep(0,nz);
  c2t=matrix(0,nrow=nz,ncol=nr);
  for(iz in 1:nz){
    c1t[iz]=-v_c1*c1z[iz]-(2/r2l)*
            (k12f*c1[iz]-k12r*c2[iz,1]);
  for(ir in 1:nr){
    c2t[iz,ir]=D2*(c2rr[iz,ir]+
            (1/r2[ir])*c2r[iz,ir]);
  }
  }
  c1t[1]=0;
```

The similarity of eqs. (1.1c,h) and this coding is one of the advantages of the MOL. Since BC (1.3) specifies $c_1(z = 0, t)$, the derivative of this value is set to zero, `c1t[1]=0`, so that `lsodes` does not change the BC specified value.

- The vector `c1t` and matrix `c2t` are placed in a single derivative vector `ut` for return to `lsodes`.

```
#
# One vector, one matrix to one vector
  ut=rep(0,(nz+nznr));
```

```
for(iz in 1:nz){
  ut[iz]=c1t[iz];
for(ir in 1:nr){
  izir=(iz-1)*nr+ir;
  ut[nz+izir]=c2t[iz,ir];
}
}
```

- The counter for the calls to pde1a is incremented and returned to the main program of Listing 2.1 with <<-.

```
#
# Increment calls to pde1a
  ncall <<- ncall+1;
```

- ut is returned to lsodes as a list (required by lsodes). c is the R vector utility.

```
#
# Return derivative vector
  return(list(c(ut)));
}
```

The final } concludes pde1a.

The output from the main program and subordinate routine of Listings 2.1, 2.2 is considered next.

(2.2.3) Model output

Abbreviated numerical output follows.

```
[1] 6
```

```
[1] 148
```

```
ncall = 1071
```

t	z	c1(z,t)	c2l(r,z,t)	c2u(r,z,t)
0.00	0.00	1.0000	0.0000	0.0000
0.00	0.10	0.0000	0.0000	0.0000
0.00	0.20	0.0000	0.0000	0.0000
0.00	0.30	0.0000	0.0000	0.0000
0.00	0.40	0.0000	0.0000	0.0000
0.00	0.50	0.0000	0.0000	0.0000
0.00	0.60	0.0000	0.0000	0.0000
0.00	0.70	0.0000	0.0000	0.0000
0.00	0.80	0.0000	0.0000	0.0000
0.00	0.90	0.0000	0.0000	0.0000
0.00	1.00	0.0000	0.0000	0.0000

t	z	c1(z,t)	c2l(r,z,t)	c2u(r,z,t)
0.40	0.00	1.0000	0.9988	0.4270
0.40	0.10	0.7325	0.7313	0.1899
0.40	0.20	0.4409	0.4399	0.0495
0.40	0.30	0.1672	0.1666	0.0032
0.40	0.40	0.0054	0.0053	0.0000
0.40	0.50	-0.0000	-0.0000	-0.0000
0.40	0.60	-0.0000	-0.0000	-0.0000
0.40	0.70	-0.0000	-0.0000	-0.0000
0.40	0.80	-0.0000	-0.0000	-0.0000
0.40	0.90	-0.0000	-0.0000	-0.0000
0.40	1.00	-0.0000	-0.0000	-0.0000

.
.
.

Output for t = 0.8,1.2,1.6 removed

.
.
.

t	z	c1(z,t)	c2l(r,z,t)	c2u(r,z,t)
2.00	0.00	1.0000	0.9999	0.9702

2.00	0.10	0.9707	0.9705	0.8880
2.00	0.20	0.9159	0.9156	0.7786
2.00	0.30	0.8309	0.8305	0.6477
2.00	0.40	0.7203	0.7198	0.5094
2.00	0.50	0.5939	0.5934	0.3773
2.00	0.60	0.4639	0.4634	0.2620
2.00	0.70	0.3418	0.3414	0.1698
2.00	0.80	0.2363	0.2360	0.1018
2.00	0.90	0.1525	0.1522	0.0559
2.00	1.00	0.0927	0.0925	0.0287

Table 2.1: Abbreviated output for eqs. (1.1c,h) to (1.4),
$$k_{12f} = k_{12r} = 1$$

We can note the following details about this output.

- The homogeneous (zero) ICs are confirmed at $t = 0$. This is an important check since if the ICs are not correct, the subsequent solution for $t > 0$ will not be correct.
- The output is for $t = 0, 0.4, \ldots, 2$ as programmed in Listing 2.1.
- The output is for $z = 0, 0.1, \ldots, 1$ as programmed in Listing 2.1.
- out, the array with the PDE solution, has dimensions out(6,147+1=148), as discussd previously. The offset +1 results from the inclusion of t as the first element in each of the 6 147-vectors.
- $c_1(z = 0, t)$ is constant according to the inlet concentration c1[1]=c1e. Thus, at $t = 0$, $c_1(z = 0, t = 0) = 1$ and at $t = 2$, $c_1(z = 0, t = 2) = 1$.
- $c_2(r = r_{2l}, z = 0, t)$ responds almost immediately as expected, e.g., $c_2(r = r_{2l}, z = 0, t = 0.4) = 0.9885$. This value is less than the corresponding value of $c_1(z = 0, t = 0.4) = 1$ since the epithelial concentration is driven by the blood concentration through BC (1.4a).

- $c_2(r = r_{2u}, z = 0, t = 0.4) = 0.4183$ follows $c_2(r = r_{2l}, z = 0, t = 0.4) = 0.9885$, for example, as the concentration in the epithelial membrane responds to $c_1(z = 0, t) = 1$
- BC (1.4b) is not obvious from the solution in Table 2.1, but this no flux or zero slope condition is apparent in the graphical output that follows.
- The computational effort to produce the solution is acceptable, `ncall = 1071`.

The graphical output follows.

The discontinuity at $z = 0$ is appproximated by a line of finite slope due to the gridding of 21 points in z. t is plotted parametrically in Figs. 2.1 for $t = 0, 0.4, \ldots, 2$ with the bottom curve (horizontal line at zero) corresponding to the homogeneous ICs (Table 2.1, $t = 0$).

The next case for the use of the main program and ODE routine of Listings 1.1 and 1.2 has two changes in the selection of the transport coefficients k_{12f}, k_{12r} and the time interval $0 \leq t \leq t_f$.

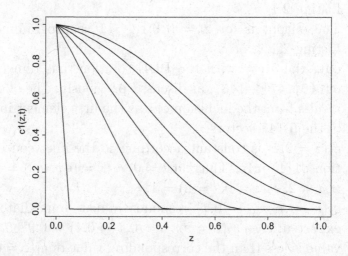

Figure 2.1a: Numerical solution $c_1(z, t)$, $k_{12f} = k_{12r} = 1$

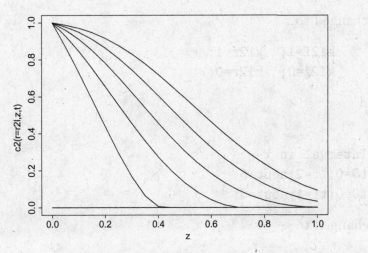

Figure 2.1b: Numerical solution $c_2(r = r_{2l}, z, t)$, $k_{12f} = k_{12r} = 1$

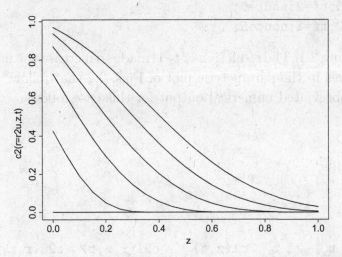

Figure 2.1c: Numerical solution $c_2(r = r_{2u}, z, t)$, $k_{12f} = k_{12r} = 1$

Specifically,

```
        k12f=1;   k12r=1;
 #      k12f=0;   k12r=0;
```

is changed to

```
#      k12f=1;   k12r=1;
       k12f=0;   k12r=0;
```

and

```
#
# Interval in t
  t0=0;tf=2;nout=6;
# t0=0;tf=1;nout=6;
```

is changed to

```
#
# Interval in t
# t0=0;tf=2;nout=6;
  t0=0;tf=1;nout=6;
```

in Listing 2.1. The reduction of t_f is made to increase the number of curves in the parameteric plot of Figs. 2.2 that follow.

Abbreviated numerical output for this case follows.

```
[1] 6

[1] 148

 ncall = 1156
```

t	z	c1(z,t)	c2l(r,z,t)	c2u(r,z,t)
0.00	0.00	1.0000	0.0000	0.0000
0.00	0.10	0.0000	0.0000	0.0000
0.00	0.20	0.0000	0.0000	0.0000
0.00	0.30	0.0000	0.0000	0.0000
0.00	0.40	0.0000	0.0000	0.0000
0.00	0.50	0.0000	0.0000	0.0000
0.00	0.60	0.0000	0.0000	0.0000

t	z	c1(z,t)	c21(r,z,t)	c2u(r,z,t)
0.00	0.70	0.0000	0.0000	0.0000
0.00	0.80	0.0000	0.0000	0.0000
0.00	0.90	0.0000	0.0000	0.0000
0.00	1.00	0.0000	0.0000	0.0000
t	z	c1(z,t)	c21(r,z,t)	c2u(r,z,t)
0.20	0.00	1.0000	0.0000	0.0000
0.20	0.10	0.9442	0.0000	0.0000
0.20	0.20	0.5520	0.0000	0.0000
0.20	0.30	0.0183	0.0000	0.0000
0.20	0.40	0.0000	0.0000	0.0000
0.20	0.50	0.0000	0.0000	0.0000
0.20	0.60	0.0000	0.0000	0.0000
0.20	0.70	0.0000	0.0000	0.0000
0.20	0.80	0.0000	0.0000	0.0000
0.20	0.90	0.0000	0.0000	0.0000
0.20	1.00	0.0000	0.0000	0.0000

.

.

.

Output for t=0.4,0.6,0.8 removed

.

.

.

t	z	c1(z,t)	c21(r,z,t)	c2u(r,z,t)
1.00	0.00	1.0000	0.0000	0.0000
1.00	0.10	1.0000	0.0000	0.0000
1.00	0.20	1.0000	0.0000	0.0000
1.00	0.30	1.0000	0.0000	0.0000
1.00	0.40	1.0000	0.0000	0.0000
1.00	0.50	1.0000	0.0000	0.0000
1.00	0.60	0.9999	0.0000	0.0000
1.00	0.70	0.9991	0.0000	0.0000
1.00	0.80	0.9901	0.0000	0.0000

1.00	0.90	0.8904	0.0000	0.0000
1.00	1.00	0.4916	0.0000	0.0000

Table 2.2: Abbreviated output for eqs. (1.1c,h) to (1.4),
$$k_{12f} = k_{12r} = 0$$

We can note the following details about this output.

- The IC is again confirmed.
- The output interval in t is 0.2 rather than 0.4 in Table 2.1 due to the change in t_f from 2 to 1.
- The computational effort is acceptable, `ncall = 1156`.

The graphical output in Figs. 2.2 reflect the solution in Table 2.2

Fig. 2.2a indicates a traveling wave according to eq. (2.1d), but also it has a non-smooth gridding effect (kinks) due to `nz=21`. Therefore, for a second case for which $c_2(r, z, t)$ remains at zero (considered subsequently), eq. (1.2h) is dropped and only `nz` MOL/ODEs are used (rather than `nz+nz*nr`). In other words, `nz` can be increased while the number of ODEs remains tractable.

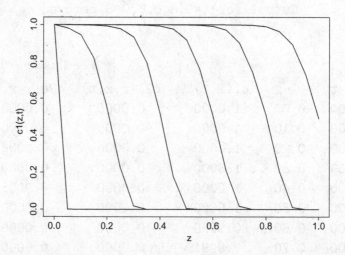

Figure 2.2a: Numerical solution $c_1(z, t)$, $k_{12f} = k_{12r} = 0$

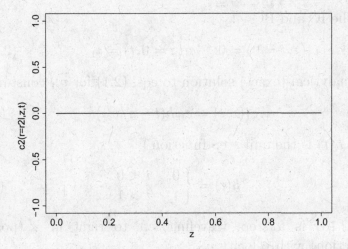

Figure 2.2b: Numerical solution $c_2(r = r_{2l}, z, t)$, $k_{12f} = k_{12r} = 0$

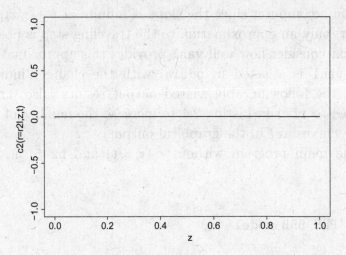

Figure 2.2c: Numerical solution $c_2(r = r_{2u}, z, t)$, $k_{12f} = k_{12r} = 0$

For this solution with $k_{12f} = k_{12r} = 0$, the transfer between the blood and membrane is then zero and the membrane concentration remains at its zero IC. In other words, eq. (1.1c) for this case is the linear advection equation

$$\frac{\partial c_1(z, t)}{\partial t} = -\frac{\partial v_{c1}(z, t) c_1(z, t)}{\partial z} \tag{2.1a}$$

with the IC and BC

$$c_1(z, t = 0) = 0; \quad c_1(z = 0, t) = c_{1e} \qquad (2.1b,c)$$

The analytical (exact) solution to eqs. (2.1) for v_{c1} constant is

$$c_1(z, t) = c_{1e}h(t - z/v_{c1}) \qquad (2.1d)$$

where $h(t)$ is the unit step function

$$h(t) = \begin{cases} 0, & t < 0 \\ 1, & t > 1 \end{cases} \qquad (2.1e)$$

Eq. (2.1d) is a step, traveling left to right in z (positive z direction) with velocity v_{c1}.

This solution is essentially impossible to reproduce numerically on a computer since the slope is infinite at $t - z/v_{c1} = 0$. Rather, only an approximation to the traveling step is possible. We next consider how well `van1` provides this approximation.

If `van1` is selected in `pde1a`, with the changes indicated above, the following abbreviated output results (also, `tf=2` is changed to `tf=1` in Listing 2.1 to increase the number of parametric curves in t in the graphical output)

The main program without $c_2(r, z, t)$ and `nz=51` is listed next.

```
#
# Two PDE BBB model
#
# Delete previous workspaces
  rm(list=ls(all=TRUE))
#
# Access ODE integrator
  library("deSolve");
#
# Access functions for numerical solution
  setwd("f:/BBB/chap2/ex1");
```

```
  source("pde1b.R");
  source("vanl.R");
#
# Model parameters
      u10=0;    u20=0;
  D2=1.0e-06;  v_c1=1;
      c1e=1;
#     k12f=1;  k12r=1;
      k12f=0;  k12r=0;
#
# Initial condition
  nz=51;
  u0=rep(0,nz);
  for(iz in 1:nz){
    u0[iz]=u10;
  }
#
# Grid in z
  zl=0;zu=1;
  z=seq(from=zl,to=zu,by=(zu-zl)/(nz-1));
#
# Interval in t
# t0=0;tf=2;nout=6;
  t0=0;tf=1;nout=6;
  tout=seq(from=t0,to=tf,by=(tf-t0)/(nout-1));
  ncall=0;
#
# ODE integration
  out=lsodes(y=u0,times=tout,func=pde1b,
      sparsetype="sparseint",rtol=1e-6,
      atol=1e-6,maxord=5);
  nrow(out)
  ncol(out)
#
```

```
# Store solution
  c1=matrix(0,nrow=nz,ncol=nout);
  t=rep(0,nout);
  for(it in 1:nout){
    t[it]=out[it,1];
    for(iz in 1:nz){
      c1[iz,it]=out[it,iz+1];
    }
  c1[1,it]=c1e;
  }
#
# Display ncall
  cat(sprintf("\n\n ncall = %2d",ncall));
#
# Display numerical solution
  for(it in 1:nout){
  cat(sprintf(
    "\n\n      t      z    c1(z,t)"));
  izv=seq(from=1,to=nz,by=5);
  for(iz in izv){
    cat(sprintf("\n%7.2f%7.2f%10.4f",
      t[it],z[iz],c1[iz,it]));
  }
  }
#
# Plot numerical solutions
#
# c1(z,t)
  par(mfrow=c(1,1));
  matplot(
    x=z,y=c1,type="l",xlab="z",ylab="c1(z,t)",
    xlim=c(zl,zu),lty=1,main="",lwd=2,col="black");
```

Listing 2.3: Main program for eq. (1.1c)

Note in particular that `nz` has been increased from 21 to 51. Also, the output is for $c_1(z, t)$ only ($c_2(r, z, t)$ is dropped).

The MOL/ODE routine `pde1b` called by `lsodes` in Listing 2.3 follows.

```
pde1b=function(t,u,parms){
#
# Function pde1b computes the t derivative
# vector of c1t(z,t)
#
# One vector to one vector
  c1=rep(0,nz);
  for(iz in 1:nz){
    c1[iz]=u[iz];
  }
#
# c1 BC
  c1[1]=c1e;
#
# c1z
  c1z=vanl(zl,zu,nz,c1,v_c1);
#
# c1t
  c1t=rep(0,nz);
  for(iz in 1:nz){
    c1t[iz]=-v_c1*c1z[iz];
  }
  c1t[1]=0;
#
# One vector to one vector
  ut=rep(0,nz);
  for(iz in 1:nz){
    ut[iz]=c1t[iz];
  }
```

```
#
# Increment calls to pde1b
  ncall<<-ncall+1;
#
# Return derivative vector
  return(list(c(ut)));
}
```

Listing 2.4: **pde1b** for eq. (1.1c)

In particular, the coding for $c_2(r, z, t)$ is removed from Listing 2.2 and the coding is for only eq. (2.1a) for constant v_{1e} (the linear advection equation).

Abbreviated numerical output follows.

```
[1] 6

[1] 52

ncall = 2848
```

t	z	c1(z,t)
0.00	0.00	1.0000
0.00	0.10	0.0000
0.00	0.20	0.0000
0.00	0.30	0.0000
0.00	0.40	0.0000
0.00	0.50	0.0000
0.00	0.60	0.0000
0.00	0.70	0.0000
0.00	0.80	0.0000
0.00	0.90	0.0000
0.00	1.00	0.0000

t	z	c1(z,t)
0.20	0.00	1.0000
0.20	0.10	0.9976
0.20	0.20	0.5424
0.20	0.30	-0.0000
0.20	0.40	-0.0000
0.20	0.50	-0.0000
0.20	0.60	-0.0000
0.20	0.70	-0.0000
0.20	0.80	-0.0000
0.20	0.90	-0.0000
0.20	1.00	-0.0000

.

.

.

Output for t = 0.4,0.6,
    0.8 removed

.

.

.

t	z	c1(z,t)
1.00	0.00	1.0000
1.00	0.10	1.0000
1.00	0.20	1.0000
1.00	0.30	1.0000
1.00	0.40	1.0000
1.00	0.50	1.0000
1.00	0.60	1.0000
1.00	0.70	1.0000
1.00	0.80	1.0000
1.00	0.90	0.9909
1.00	1.00	0.4212

Table 2.3: Abbreviated output for eq. (1.1c), $k_{12f} = k_{12r} = 0$

We can note the following details about this output.

- The dimensions of out (from lsodes) are out(6,51+ 1=52) as expected (since nz=51).

  [1] 6

  [1] 52

- The output for t is for an interval of 0.2 that reflects the programming in Listing 2.3 (with $t_f = 1$). In particular, the BC $c_1(z = 0, t) = 1$ is confirmed for $t = 0$, 0.2, 1.
- The output is for an interval of 0.1 in z from by=5 in the output statement in Listing 2.3.
- The traveling wave (step) solution of eq. (2.1d) is reflected in the successive values of $c_1(z, t)$, particularly for $t = z/v_{1e} = 0$ where the discontuinuity of eq. (2.1d) occurs.
- The computational effort is acceptable, ncall = 2848.

The graphical output is in Fig. 2.3.

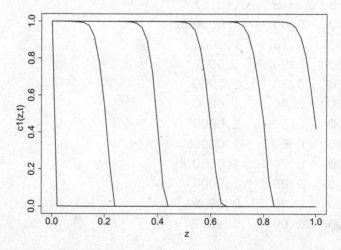

Figure 2.3: Numerical solution $c_1(z, t)$, $k_{12f} = k_{12r} = 0$, $c_2(r, z, t)$ removed

$c_1(z, t)$ corresponds to the traveling wave solution of eq. (2.1d).

This completes the discussion of the two PDE model of eqs. (1.1c,h) to (1.4).

(2.3) Summary and Conclusions

In this chapter, a 2×2 PDE system to model the transfer between the BBB capillary blood flow and the epithelial membrane is presented, including the computer implementation of the model.

Two types of spatial derivatives are considered in eqs. (1.1c) and (1.1h).

1. Eq. (1.1c) is termed a first order, hyperbolic PDE. It can propagate steep fronts and discontinuities so computing a numerical solution can be difficult. In particular, the z derivative $\dfrac{\partial c_1(z, t)}{\partial z}$ in eq. (1.1c) requires special consideration, which in the present case is reflected in use of the van Leer flux limiter in vanl (listed in Appendix A3). If this derivative is computed with FDs, the solution will have errors in the form of numerical diffusion and oscillation [2]. vanl gives no oscillations and limited diffusion (dispersion) around the moving front (step), as reflected in Fig. 2.3.

   Two flux limiters are listed in Appendix A4 that can be used in a comparative study with vanl in Appendix A3 (the calling sequences for the three flux limiters are the same and only the name of the routines is different).

2. Eq. (1.1h) is termed a parabolic PDE. It models diffusion (in this case in the membrane) and therefore tends to smooth discontinuities so that computing a solution is easier than for a hyperbolic PDE. In other words, FDs can be used, as in dss004, dss044.

Eqs. (1.1c) and (1.1h) therefore constitute a hyperbolic-parabolic (convective-diffusive) PDE system to which the numerical methods in Listings 2.1 to 2.4 can be applied.

In the next chapter, a three PDE model is considered which also includes the brain tissue. That is, the two PDE model in this chapter is extended with the addition of a PDE for the brain tissue.

References

[1] Soetaert, K., J. Cash, and F. Mazzia (2012), *Solving Differential Equations in R*, Springer-Verlag, Heidelberg, Germany.

[2] Griffiths, G.W., and W.E. Schiesser (2012), *Traveling Wave Analysis of Partial Diffenetial Equations*, Elsevier, Burlington, MA, USA.

Chapter 3

Three PDE Models of Blood Brain Transfer

(3.1) Introduction

The two PDE model of Chapter 2 is now extended to 3×3 by adding a PDE for the brain tissue concentration with dependent variable $c_3(r, z, t)$ (this extension demonstrates how a PDE can be added to the BBB/PDE model). Fig. 1.1 is a schematic diagram for the three PDE system.

(3.2) Three PDE Model

The following eqs. (3.1) constitute the three PDE model.

$$\frac{\partial c_1(z,t)}{\partial t} = -v_{c1}\frac{\partial c_1(z,t)}{\partial z} - \frac{2}{r_1}(k_{12f}c_1(z,t) - k_{12r}c_2(r = r_1, z, t))$$

$$\text{(3.1a)}$$

$$\frac{\partial c_2(r,z,t)}{\partial t} = D_2\left(\frac{\partial^2 c_2(r,z,t)}{\partial r^2} + \frac{1}{r}\frac{\partial c_2(r,z,t)}{\partial r}\right) \quad \text{(3.1b)}$$

$$\frac{\partial c_3(r,z,t)}{\partial t} = D_3\left(\frac{\partial^2 c_3(r,z,t)}{\partial r^2} + \frac{1}{r}\frac{\partial c_3(r,z,t)}{\partial r}\right) - k_1 c_3(r,z,t)$$

$$\text{(3.1c)}$$

Eq. (3.1a) is the same as eq. (1.1c). Eq. (3.1b) is eq. (1.1h) (for constant D_2). Eq. (3.1c) is a mass balance for the tissue that is derived in the same way as eq. (1.1h), with the addition of a volumetric reaction rate, $-k_1 c_3(r, z, t)$.

43

Eq. (3.1a) requires one IC and one BC.

$$c_1(z, t = 0) = c_{10}(z); \ c_1(z = 0, t) = c_{1e}(t) \qquad \text{(3.2a,b)}$$

Eq. (3.1b) requires one IC and two BCs.

$$c_2(r, z, t = 0) = c_{20}(r, z) \qquad \text{(3.3a)}$$

$$D_2 \frac{c_2(r = r_{2l}, z, t)}{\partial r} = -(k_{12f}c_1(z, t) - k_{12r}c_2(r = r_{2l}, z, t))$$

$$\text{(3.3b)}$$

$$D_2 \frac{c_2(r = r_{2u}, z, t)}{\partial r} = -(k_{23f}c_2(r = r_{2u}, z, t) - k_{23r}c_3(r = r_{2u}, z, t))$$

$$\text{(3.3c)}$$

Eqs. (3.3b,c) equate the mass transfer rates across the lower and upper boundaries of the membrane, respectively. The transfer coefficients for the blood to the lower membrane boundary are k_{12f}, k_{12r} and for the upper membrane boundary to lower tissue boundary they are k_{23f}, k_{23r} (see Fig. 1.1).

Eq. (3.1c) requires one IC and two BCs.

$$c_3(r, z, t = 0) = c_{30}(r, z) \qquad \text{(3.4a)}$$

$$D_3 \frac{c_3(r = r_{3l}, z, t)}{\partial r} = -(k_{23f}c_2(r = r_{2u}, z, t) - k_{23r}c_3(r = r_{3l}, z, t))$$

$$\text{(3.4b)}$$

$$D_3 \frac{c_3(r = r_{3u}, z, t)}{\partial r} = 0 \qquad \text{(3.4c)}$$

Eq. (3.4b) equates the mass transfer rates at the lower boundary of the tissue in terms of the mass transfer coefficients k_{23f}, k_{23r}. Eq. (3.4c) is a no flux BC at the upper boundary of the tissue.

Eqs. (3.1) to (3.4) are a 3×3 PDE system. The programming is discussed next starting with a main program.

(3.2.1) Main program

The main program for eqs. (3.1) to (3.4) follows.

```
#
#  Three PDE BBB model
#
# Delete previous workspaces
  rm(list=ls(all=TRUE))
#
# Access ODE integrator
  library("deSolve");
#
# Access functions for numerical solution
  setwd("f:/BBB/chap3");
  source("pde1a.R");
  source("dss004.R");
  source("dss044.R");
  source("van1.R");
#
# Model parameters
      u10=0;       u20=0;
      u30=0;      v_c1=1;
      c1e=1;        k1=1;
  D2=1.0e-06; D3=1.0e-06;
     k12f=1;      k12r=1;
     k23f=1;      k23r=1;
#    k12f=0;      k12r=0;
#    k23f=0;      k23r=0;
#
# Initial condition
  nz=21;nr=6;
  nznr=nz*nr;
  u0=rep(0,nz+2*nznr);
  for(iz in 1:nz){
```

```
   u0[iz]=u10;
  for(ir in 1:nr){
    izir=(iz-1)*nr+ir;
    u0[nz+izir]      =u20;
    u0[nz+izir+nznr]=u30;
  }
  }
#
# Grid in z
  zl=0;zu=1;
  z=seq(from=zl,to=zu,by=(zu-zl)/(nz-1));
#
# Grid in r for c2
  r2l=1.0e-03;r2u=2.0e-03;
  r2=seq(from=r2l,to=r2u,by=(r2u-r2l)/(nr-1));
#
# Grid in r for c3
  r3l=2.0e-03;r3u=3.0e-03;
  r3=seq(from=r3l,to=r3u,by=(r3u-r3l)/(nr-1));
#
# Interval in t
  t0=0;tf=2;nout=6;
# t0=0;tf=1;nout=6;
  tout=seq(from=t0,to=tf,by=(tf-t0)/(nout-1));
  ncall=0;
#
# ODE integration
  out=lsodes(y=u0,times=tout,func=pde1a,
      sparsetype="sparseint",rtol=1e-6,
      atol=1e-6,maxord=5);
  nrow(out)
  ncol(out)
#
# Store solution
```

```
c1 =matrix(0,nrow=nz,ncol=nout);
c2l=matrix(0,nrow=nz,ncol=nout);
c2u=matrix(0,nrow=nz,ncol=nout);
c3l=matrix(0,nrow=nz,ncol=nout);
c3u=matrix(0,nrow=nz,ncol=nout);
t=rep(0,nout);
for(it in 1:nout){
for(iz in 1:nz){
   c1[iz,it]=out[it,iz+1];
  c2l[iz,it]=out[it,nz+1+(iz-1)*nr+1];
  c2u[iz,it]=out[it,nz+1+(iz-1)*nr+nr];
  c3l[iz,it]=out[it,nz+1+(iz-1)*nr+1+nznr];
  c3u[iz,it]=out[it,nz+1+(iz-1)*nr+nr+nznr];
       t[it]=out[it,1];
}
c1[1,it]=c1e;
}
#
# Display ncall
  cat(sprintf("\n\n ncall = %2d",ncall));
#
# Display numerical solution
  for(it in 1:nout){
  cat(sprintf(
    "\n\n      t      z     c1(z,t)      c2l(r,z,t)
    c2u(r,z,t)"));
  cat(sprintf(
    "\n      t      z     c1(z,t)      c3l(r,z,t)
    c3u(r,z,t)"));
  izv=seq(from=1,to=nz,by=2);
  for(iz in izv){
    cat(sprintf(
      "\n%7.2f%7.2f%10.4f%12.4f%12.4f",
      t[it],z[iz],c1[iz,it],c2l[iz,it],
```

```
      c2u[iz,it])));
    cat(sprintf(
      "\n%7.2f%7.2f%10.4f%12.4f%12.4f\n",
      t[it],z[iz],c1[iz,it],c3l[iz,it],
      c3u[iz,it])));
  }
}
#
# Plot numerical solutions
#
# c1(z,t)
  par(mfrow=c(1,1));
  matplot(
    x=z,y=c1,type="l",xlab="z",
    ylab="c1(z,t)",xlim=c(zl,zu),
    lty=1,main="",lwd=2,col="black");
#
# c2(r=r2l,z,t)
  par(mfrow=c(1,1));
  matplot(
    x=z,y=c2l,type="l",xlab="z",
    ylab="c2(r=r2l,z,t)",xlim=c(zl,zu),
    lty=1,main="",lwd=2,col="black");
#
# c2(r=r2u,z,t)
  par(mfrow=c(1,1));
  matplot(
    x=z,y=c2u,type="l",xlab="z",
    ylab="c2(r=r2u,z,t)",xlim=c(zl,zu),
    lty=1,main="",lwd=2,col="black");
#
# c3(r=rl3,z,t)
  par(mfrow=c(1,1));
  matplot(
```

```
    x=z,y=c3l,type="l",xlab="z",
    ylab="c3(r=r3l,z,t)",xlim=c(zl,zu),
    lty=1,main="",lwd=2,col="black");
#
# c3(r=ru3,z,t)
  par(mfrow=c(1,1));
  matplot(
    x=z,y=c3u,type="l",xlab="z",
    ylab="c3(r=r3u,z,t)",xlim=c(zl,zu),
    lty=1,main="",lwd=2,col="black");
```

Listing 3.1: Main program for eqs. (3.1) to (3.4)

We can note the following details about Listing 3.1.

- Previous workspaces are deleted.

```
#
# Three PDE BBB model
#
# Delete previous workspaces
  rm(list=ls(all=TRUE))
```

- The R ODE integrator library deSolve is accessed. Then the directory with the files for the solution of eqs. (3.1) to (3.4) is designated. Note that setwd (set working directory) uses / rather than the usual \.

```
#
# Access functions for numerical solution
  setwd("f:/BBB/chap3");
  source("pde1a.R");
  source("dss004.R");
  source("dss044.R");
  source("van1.R");
```

pde1a.R is the routine for the method of lines (MOL) approximation of PDEs (3.1) (discussed subsequently).

dss004, dss044 (Differentiation in Space Subroutine) are library routines for calculating first and second derivatives in r by finite differences (FDs). vanl is a library routine for calculating first derivatives in z. The coding and use of these routines is discussed subsequently.

- The model parameters are defined numerically.

```
#
# Model parameters
      u10=0;        u20=0;
      u30=0;        v_c1=1;
      c1e=1;        k1=1;
  D2=1.0e-06;  D3=1.0e-06;
      k12f=1;      k12r=1;
      k23f=1;      k23r=1;
  #   k12f=0;      k12r=0;
  #   k23f=0;      k23r=0;
```

Two cases are programmed. The second case can be selected by removing the comment chracter, #.

- The initial conditions (ICs) for eqs. (3.1) are placed in a vector u0 of length nz + 2*nz*nr = 21 + 2*21*6 = 273.

```
#
# Initial condition
  nz=21;nr=6;
  nznr=nz*nr;
  u0=rep(0,nz+2*nznr);
  for(iz in 1:nz){
    u0[iz]=u10;
    for(ir in 1:nr){
      izir=(iz-1)*nr+ir;
      u0[nz+izir]      =u20;
```

```
    u0[nz+izir+nznr]=u30;
  }
}
```

- A spatial grid of nz=21 points is defined for $z_l = 0 \leq z \leq z_u = 1$, so that $z = 0, 0.05, \ldots, 1$.

```
#
# Grid in z
  zl=0;zu=1;
  z=seq(from=zl,to=zu,by=(zu-zl)/(nz-1));
```

- A spatial grid of 6 points is defined for the membrane, $r_{2l} = 0.001 \leq r_2 \leq r_{2u} = 0.002$, so that $r_2 = 0.001$, $0.0012, \ldots, 0.002$.

```
#
# Grid in r for c2
  r2l=1.0e-03;r2u=2.0e-03;
  r2=seq(from=r2l,to=r2u,by=(r2u-r2l)/(nr-1));
```

These dimensions demonstrate that the computer implementation of eqs. (3.1) to (3.4) will execute with small values corresponding to representative BBB dimensions.

- A spatial grid of 6 points is defined for the tissue, $r_{3l} = 0.002 \leq r_3 \leq r_{3u} = 0.003$, so that $r_3 = 0.002$, $0.0022, \ldots, 0.003$.

```
#
# Grid in r for c3
  r3l=2.0e-03;r3u=3.0e-03;
  r3=seq(from=r3l,to=r3u,by=(r3u-r3l)/(nr-1));
```

- An interval in t of 6 points is defined for $0 \leq t \leq 2$ so that tout=0,0.4,...,2.

```
#
# Interval in t
  t0=0;tf=2;nout=6;
```

```
# t0=0;tf=1;nout=6;
  tout=seq(from=t0,to=tf,by=(tf-t0)/(nout-1));
  ncall=0;
```

A second case with `tf=1` is considered subsequently. The counter for the calls to the ODE/MOL routine `pde1a` is also initialized.

- The system of 273 MOL/ODEs is integrated by the library integrator `lsodes` (available in `deSolve`). As expected, the inputs to `lsodes` are the ODE function, `pde1a`, the IC vector u0, and the vector of output values of t, `tout`. The length of u0 (e.g., 273) informs `lsodes` how many ODEs are to be integrated. `func,y,times` are reserved names.

```
#
# ODE integration
  out=lsodes(y=u0,times=tout,func=pde1a,
      sparsetype="sparseint",rtol=1e-6,
      atol=1e-6,maxord=5);
  nrow(out)
  ncol(out)
```

The numerical solution to the ODEs is returned in matrix `out`. In this case, `out` has the dimensions $nout \times (nz + 2(nz)(nr) + 1) = 6 \times 21 + (2)(21)(6) + 1 = 274$, which are confirmed by the output from `nrow(out),ncol(out)` (included in the numerical output considered subsequently).

The offset $273 + 1 = 274$ is required since the first element of each column has the output t (also in `tout`), and the $2, \ldots, 273 + 1 = 2, \ldots, 274$ column elements have the 273 ODE solutions.

- Selected solutions of the 273 ODEs returned in `out` by `lsodes` are placed in arrays `c1`, `c21`, `c2u`, `c31`, `c3u`.

```
#
# Store solution
  c1 =matrix(0,nrow=nz,ncol=nout);
  c2l=matrix(0,nrow=nz,ncol=nout);
  c2u=matrix(0,nrow=nz,ncol=nout);
  c3l=matrix(0,nrow=nz,ncol=nout);
  c3u=matrix(0,nrow=nz,ncol=nout);
  t=rep(0,nout);
  for(it in 1:nout){
  for(iz in 1:nz){
    c1[iz,it]=out[it,iz+1];
    c2l[iz,it]=out[it,nz+1+(iz-1)*nr+1];
    c2u[iz,it]=out[it,nz+1+(iz-1)*nr+nr];
    c3l[iz,it]=out[it,nz+1+(iz-1)*nr+1+nznr];
    c3u[iz,it]=out[it,nz+1+(iz-1)*nr+nr+nznr];
        t[it]=out[it,1];
  }
  c1[1,it]=c1e;
  }
```

Again, the offset +1 is required since the first element of each column of out has the value of t. $c_1(z = 0, t) =$ c1[1,it]=c1e sets BC (3.2b) since this value is not returned by lsodes (it is defined algebraically in pde1a and not by the integration of an ODE at $z = 0$).
$c_2(r = r_{2l}, z, t), c_2(r = r_{2u}, z, t)$ are placed in c2l, c2u.
$c_3(r = r_{3l}, z, t), c_3(r = r_{3u}, z, t)$ are placed in c3l, c3u.
- The number of calls to pde1a is displayed at the end of the solution.

```
#
# Display ncall
  cat(sprintf("\n\n ncall = %2d",ncall));
```

- $t, z, c_1(z,t)$, $c_2(r = r_{2l}, z, t), c_2(r = r_{2u}, z, t)$, $c_3(r = r_{3l}, z, t), c_3(r = r_{3u}, z, t)$ are displayed for every second value of z from by=2.

```
#
# Display numerical solution
  for(it in 1:nout){
  cat(sprintf(
    "\n\n        t        z    c1(z,t)        c21(r,z,t)
      c2u(r,z,t)"));
  cat(sprintf(
     "\n        t        z    c1(z,t)        c31(r,z,t)
       c3u(r,z,t)"));
  izv=seq(from=1,to=nz,by=2);
  for(iz in izv){
    cat(sprintf(
     "\n%7.2f%7.2f%10.4f%12.4f%12.4f",
     t[it],z[iz],c1[iz,it],c21[iz,it],
     c2u[iz,it]));
    cat(sprintf(
     "\n%7.2f%7.2f%10.4f%12.4f%12.4f\n",
     t[it],z[iz],c1[iz,it],c31[iz,it],
     c3u[iz,it]));
  }
  }
```

- Five selected values of the solution are plotted against z and parametrically in t.

```
#
# Plot numerical solutions
#
# c1(z,t)
  par(mfrow=c(1,1));
```

```
    matplot(
      x=z,y=c1,type="l",xlab="z",
      ylab="c1(z,t)",xlim=c(zl,zu),
      lty=1,main="",lwd=2,col="black");
#
# c2(r=r2l,z,t)
  par(mfrow=c(1,1));
  matplot(
    x=z,y=c2l,type="l",xlab="z",
    ylab="c2(r=r2l,z,t)",xlim=c(zl,zu),
    lty=1,main="",lwd=2,col="black");
#
# c2(r=r2u,z,t)
  par(mfrow=c(1,1));
  matplot(
    x=z,y=c2u,type="l",xlab="z",
    ylab="c2(r=r2u,z,t)",xlim=c(zl,zu),
    lty=1,main="",lwd=2,col="black");
#
# c3(r=rl3,z,t)
  par(mfrow=c(1,1));
  matplot(
    x=z,y=c3l,type="l",xlab="z",
    ylab="c3(r=r3l,z,t)",xlim=c(zl,zu),
    lty=1,main="",lwd=2,col="black");
#
# c3(r=ru3,z,t)
  par(mfrow=c(1,1));
  matplot(
    x=z,y=c3u,type="l",xlab="z",
    ylab="c3(r=r3u,z,t)",xlim=c(zl,zu),
    lty=1,main="",lwd=2,col="black");
```

This completes the main program. The ODE/MOL routine pde1a called by lsodes is considered next.

(3.2.2) ODE/MOL routine

pde1a follows.

```
  pde1a=function(t,u,parms){
#
# Function pde1a computes the t derivative
# vector of c1t(z,t), c2t(r,z,t), c3t(r,z,t)
#
# One vector to one vector, two matrices
  c1=rep(0,nz);
  c2=matrix(0,nrow=nz,ncol=nr);
  c3=matrix(0,nrow=nz,ncol=nr);
  for(iz in 1:nz){
    c1[iz]=u[iz];
  for(ir in 1:nr){
    izir=(iz-1)*nr+ir;
    c2[iz,ir]=u[izir+nz];
    c3[iz,ir]=u[izir+nz+nznr];
  }
  }
#
# Boundary condition
  c1[1]=c1e;
#
# c1z
  c1z=van1(zl,zu,nz,c1,v_c1);
#
# c2r
  c2r=matrix(0,nrow=nz,ncol=nr);
  for(iz in 1:nz){
```

```
    c2r[iz,]=dss004(r21,r2u,nr,c2[iz,]);
  }
#
# c3r
  c3r=matrix(0,nrow=nz,ncol=nr);
  for(iz in 1:nz){
    c3r[iz,]=dss004(r31,r3u,nr,c3[iz,]);
  }
#
# c2 BCs
  for(iz in 1:nz){
    c2r[iz,1] =-(1/D2)*(k12f*c1[iz]-
               k12r*c2[iz,1]);
    c2r[iz,nr]=-(1/D2)*(k23f*c2[iz,nr]-
               k23r*c3[iz,1]);
  }
#
# c3 BCs
  for(iz in 1:nz){
    c3r[iz,1] =-(1/D3)*(k23f*c2[iz,nr]-
               k23r*c3[iz,1]);
    c3r[iz,nr]=0;
  }
#
# c2rr
  nl=2;nu=2;
  c2rr=matrix(0,nrow=nz,ncol=nr);
  for(iz in 1:nz){
    c2rr[iz,]=dss044(r21,r2u,nr,c2[iz,],
                     c2r[iz,],nl,nu);
  }
#
# c3rr
  nl=2;nu=2;
```

```
   c3rr=matrix(0,nrow=nz,ncol=nr);
   for(iz in 1:nz){
     c3rr[iz,]=dss044(r3l,r3u,nr,c3[iz,],
                      c3r[iz,],nl,nu);
   }
#
# c1t, c2t, c3t
   c1t=rep(0,nz);
   c2t=matrix(0,nrow=nz,ncol=nr);
   c3t=matrix(0,nrow=nz,ncol=nr);
   for(iz in 1:nz){
     c1t[iz]=-v_c1*c1z[iz]-(2/r21)*
             (k12f*c1[iz]-k12r*c2[iz,1]);
   for(ir in 1:nr){
     c2t[iz,ir]=D2*(c2rr[iz,ir]+(1/r2[ir])*
                c2r[iz,ir]);
     c3t[iz,ir]=D3*(c3rr[iz,ir]+(1/r3[ir])*
                c3r[iz,ir])-k1*c3[iz,ir];
   }
   }
   c1t[1]=0;
#
# One vector, two matrices to one vector
   ut=rep(0,nz+2*nznr);
   for(iz in 1:nz){
     ut[iz]=c1t[iz];
   for(ir in 1:nr){
     izir=(iz-1)*nr+ir;
     ut[nz+izir]     =c2t[iz,ir];
     ut[nz+izir+nznr]=c3t[iz,ir];
   }
   }
#
# Increment calls to pde1a
```

```
  ncall<<-ncall+1;
#
# Return derivative vector
  return(list(c(ut)));
}
```

Listing 3.2: ODE/MOL routine for eqs. (3.1) to (3.4)

We can note the following details about Listing 3.2.

- The function is defined.

  ```
    pde1a=function(t,u,parms){
  #
  # Function pde1a computes the t derivative
  # vector of c1t(z,t), c2t(r,z,t), c3t(r,z,t)
  ```

  t is the current value of t in eqs. (3.1). u is the 273-vector of ODE/MOL dependent variables. parm is an argument to pass parameters to pde1a (unused, but required in the argument list). The arguments must be listed in the order stated to properly interface with lsodes called in the main program of Listing 3.1. The derivative vector of the LHS of eqs. (3.1) is calculated next and returned to lsodes.

- u is placed in one vector, c1, and two matrices, c2, c3, to facilitate the programming of eqs. (3.1) to (3.4).

  ```
  #
  # One vector to one vector, two matrices
    c1=rep(0,nz);
    c2=matrix(0,nrow=nz,ncol=nr);
    c3=matrix(0,nrow=nz,ncol=nr);
    for(iz in 1:nz){
      c1[iz]=u[iz];
      for(ir in 1:nr){
  ```

```
    izir=(iz-1)*nr+ir;
    c2[iz,ir]=u[izir+nz];
    c3[iz,ir]=u[izir+nz+nznr];
  }
}
```

The dimensions for the vectors and matrices are `u(273)`, `c1(21)`, `c2(21,6)`, `c3(21,6)`.

- BC (3.2b) is implemented.

```
#
# Boundary condition
  c1[1]=c1e;
```

- $\dfrac{\partial c_1(z,t)}{\partial z}$ in eq. (3.1a) is computed by **van1**.

```
#
# c1z
  c1z=van1(zl,zu,nz,c1,v_c1);
```

`c1z` does not have to be allocated (with **rep**) since this is done by **van1**.

van1 is an implementation of the van Leer flux limiter that is designed specifically to approximate first order, convective derivatives, e.g., $\dfrac{\partial c_1(z,t)}{\partial z}$ in eq. (3.1a). **van1** is listed in Appendix A3.

- The first derivative $\dfrac{\partial c_2(r,z,t)}{\partial r}$ in eq. (3.1b) is computed by **dss004**.

```
#
# c2r
  c2r=matrix(0,nrow=nz,ncol=nr);
  for(iz in 1:nz){
    c2r[iz,]=dss004(r2l,r2u,nr,c2[iz,]);
  }
```

The values of z are selected with the `for`. The 6 values of r are selected with the , subscript. Therefore, `c2r` is a 6-vector at a particular z.

- The first derivative $\dfrac{\partial c_3(r, z, t)}{\partial r}$ in eq. (3.1c) is computed by `dss004`.

```
  }
#
# c3r
  c3r=matrix(0,nrow=nz,ncol=nr);
  for(iz in 1:nz){
    c3r[iz,]=dss004(r3l,r3u,nr,c3[iz,]);
  }
```

`c3r` is a 6-vector at a particular z.

- BCs (3.3b,c) are implemented (the subscripts `1,nr` correspond to `r2l,r2u`) at each z.

```
#
# c2 BCs
  for(iz in 1:nz){
    c2r[iz,1] =-(1/D2)*(k12f*c1[iz]-
              k12r*c2[iz,1]);
    c2r[iz,nr]=-(1/D2)*(k23f*c2[iz,nr]-
              k23r*c3[iz,1]);
  }
```

- BCs (3.4b,c) are implemented (the subscripts `1,nr` correspond to `r3l,r3u`) at each z.

```
#
# c3 BCs
  for(iz in 1:nz){
    c3r[iz,1] =-(1/D3)*(k23f*c2[iz,nr]-
              k23r*c3[iz,1]);
    c3r[iz,nr]=0;
  }
```

- The second derivative $\dfrac{\partial^2 c_2(r,z,t)}{\partial r^2}$ in eq. (3.1b) is computed by dss044.

```
#
# c2rr
  nl=2;nu=2;
  c2rr=matrix(0,nrow=nz,ncol=nr);
  for(iz in 1:nz){
    c2rr[iz,]=dss044(r2l,r2u,nr,c2[iz,],
                     c2r[iz,],nl,nu);
  }
```

Since BCs (3.3b,c) are Neumann, nl=2;nu=2 is used before calling dss044. Additional details about dss044 are available in Appendix A2.

- The second derivative $\dfrac{\partial^2 c_3(r,z,t)}{\partial r^2}$ in eq. (3.1c) is computed by dss044.

```
#
# c3rr
  nl=2;nu=2;
  c3rr=matrix(0,nrow=nz,ncol=nr);
  for(iz in 1:nz){
    c3rr[iz,]=dss044(r3l,r3u,nr,c3[iz,],
                     c3r[iz,],nl,nu);
  }
```

Since BCs (3.4b,c) are Neumann, nl=2;nu=2 is used before calling dss044.

- PDEs (3.1) are programmed, and the derivatives in t are placed in c1t, c2t, c3t.

```
#
# c1t, c2t, c3t
  c1t=rep(0,nz);
```

```
c2t=matrix(0,nrow=nz,ncol=nr);
c3t=matrix(0,nrow=nz,ncol=nr);
for(iz in 1:nz){
  c1t[iz]=-v_c1*c1z[iz]-(2/r21)*
          (k12f*c1[iz]-k12r*c2[iz,1]);
  for(ir in 1:nr){
    c2t[iz,ir]=D2*(c2rr[iz,ir]+(1/r2[ir])*
              c2r[iz,ir]);
    c3t[iz,ir]=D3*(c3rr[iz,ir]+(1/r3[ir])*
              c3r[iz,ir])-k1*c3[iz,ir];
  }
}
c1t[1]=0;
```

The similarity of eqs. (3.1) and this coding is one of the advantages of the MOL. Since BC (3.2b) specifies $c_1(z = 0, t)$, the derivative of this value is set to zero, c1t[1]=0, so that lsodes does not change the BC specified value.

- The vector c1t and matrices c2t, c3t are placed in a single derivative vector ut for return to lsodes.

```
#
# One vector, two matrices to one vector
  ut=rep(0,nz+2*nznr);
  for(iz in 1:nz){
    ut[iz]=c1t[iz];
    for(ir in 1:nr){
    izir=(iz-1)*nr+ir;
    ut[nz+izir]     =c2t[iz,ir];
    ut[nz+izir+nznr]=c3t[iz,ir];
  }
}
```

- The counter for the calls to pde1a is incremented and returned to the main program of Listing 3.1 with <<-.

```
#
# Increment calls to pde1a
  ncall<<-ncall+1;
```

- ut is returned to lsodes as a list (required by lsodes). c is the R vector utility.

```
#
# Return derivative vector
  return(list(c(ut)));
}
```

The final } concludes pde1a.

The output from the main program and subordinate routine of Listings 3.1, 3.2 is considered next.

(3.2.3) Model output

Abbreviated numerical output follows.

```
[1] 6
```

```
[1] 274
```

```
ncall = 1114
```

t	z	c1(z,t)	c21(r,z,t)	c2u(r,z,t)
t	z	c1(z,t)	c31(r,z,t)	c3u(r,z,t)
0.00	0.00	1.0000	0.0000	0.0000
0.00	0.00	1.0000	0.0000	0.0000
0.00	0.10	0.0000	0.0000	0.0000
0.00	0.10	0.0000	0.0000	0.0000
0.00	0.20	0.0000	0.0000	0.0000
0.00	0.20	0.0000	0.0000	0.0000

t	z	c1(z,t)	c2l(r,z,t)	c2u(r,z,t)
0.00	0.30	0.0000	0.0000	0.0000
0.00	0.30	0.0000	0.0000	0.0000
0.00	0.40	0.0000	0.0000	0.0000
0.00	0.40	0.0000	0.0000	0.0000
0.00	0.50	0.0000	0.0000	0.0000
0.00	0.50	0.0000	0.0000	0.0000
0.00	0.60	0.0000	0.0000	0.0000
0.00	0.60	0.0000	0.0000	0.0000
0.00	0.70	0.0000	0.0000	0.0000
0.00	0.70	0.0000	0.0000	0.0000
0.00	0.80	0.0000	0.0000	0.0000
0.00	0.80	0.0000	0.0000	0.0000
0.00	0.90	0.0000	0.0000	0.0000
0.00	0.90	0.0000	0.0000	0.0000
0.00	1.00	0.0000	0.0000	0.0000
0.00	1.00	0.0000	0.0000	0.0000

t	z	c1(z,t)	c2l(r,z,t)	c2u(r,z,t)
t	z	c1(z,t)	c3l(r,z,t)	c3u(r,z,t)
0.40	0.00	1.0000	0.9987	0.1832
0.40	0.00	1.0000	0.1828	0.0265
0.40	0.10	0.7256	0.7243	0.0841
0.40	0.10	0.7256	0.0839	0.0066
0.40	0.20	0.4391	0.4381	0.0228

0.40	0.20	0.4391	0.0227	0.0006
0.40	0.30	0.1682	0.1676	0.0021
0.40	0.30	0.1682	0.0021	0.0000
0.40	0.40	0.0099	0.0099	0.0000
0.40	0.40	0.0099	0.0000	0.0000
0.40	0.50	-0.0000	-0.0000	-0.0000
0.40	0.50	-0.0000	-0.0000	-0.0000
0.40	0.60	-0.0000	-0.0000	-0.0000
0.40	0.60	-0.0000	-0.0000	0.0000
0.40	0.70	-0.0000	-0.0000	-0.0000
0.40	0.70	-0.0000	-0.0000	0.0000
0.40	0.80	-0.0000	-0.0000	0.0000
0.40	0.80	-0.0000	0.0000	0.0000
0.40	0.90	0.0000	0.0000	0.0000
0.40	0.90	0.0000	0.0000	0.0000
0.40	1.00	0.0000	0.0000	0.0000
0.40	1.00	0.0000	0.0000	0.0000

.
.
.

Output for t=0.8,1.2,1.6 removed

.
.
.

t	z	$c1(z,t)$	$c21(r,z,t)$	$c2u(r,z,t)$
t	z	$c1(z,t)$	$c31(r,z,t)$	$c3u(r,z,t)$

2.00	0.00	1.0000	0.9991	0.4090
2.00	0.00	1.0000	0.4086	0.2333
2.00	0.10	0.8405	0.8398	0.3300
2.00	0.10	0.8405	0.3296	0.1819
2.00	0.20	0.6969	0.6962	0.2601
2.00	0.20	0.6969	0.2599	0.1374
2.00	0.30	0.5701	0.5695	0.2005
2.00	0.30	0.5701	0.2003	0.1004
2.00	0.40	0.4588	0.4584	0.1505
2.00	0.40	0.4588	0.1503	0.0707
2.00	0.50	0.3620	0.3616	0.1094
2.00	0.50	0.3620	0.1092	0.0475
2.00	0.60	0.2786	0.2783	0.0764
2.00	0.60	0.2786	0.0763	0.0301
2.00	0.70	0.2080	0.2078	0.0510
2.00	0.70	0.2080	0.0509	0.0179
2.00	0.80	0.1495	0.1493	0.0321
2.00	0.80	0.1495	0.0321	0.0097
2.00	0.90	0.1024	0.1022	0.0188
2.00	0.90	0.1024	0.0188	0.0048
2.00	1.00	0.0667	0.0666	0.0103
2.00	1.00	0.0667	0.0103	0.0022

Table 3.1: Output for eqs. (3.1) to (3.4),
$$k_{12f} = k_{12r} = k_{23f} = k_{23r} = 1$$

We can note the following details about this output.

- The homogeneous (zero) ICs are confirmed at $t = 0$. This is an important check since if the ICs are not correct, the subsequent solution for $t > 0$ will not be correct.
- The output is for $t = 0, 0.4, \ldots, 2$ as programmed in Listing 3.1.
- The output is for $z = 0, 0.1, \ldots, 1$ as programmed in Listing 3.1.
- out, the array with the PDE solution, has dimensions out(6,273+1=274), as discussd previously. The offset +1 results from the inclusion of t as the first element in each of the 6 273-vectors.
- $c_1(z = 0, t)$ is constant according to the inlet concentration c1[1]=c1e. Thus, at $t = 0$, $c_1(z = 0, t = 0) = 1$ and at $t = 2$, $c_1(z = 0, t = 2) = 1$.
- $c_2(r = r_{2l}, z = 0, t)$ responds almost immediately as expected, e.g., $c_2(r = r_{2l}, z = 0, t = 0.4) = 0.9987$. This value is less than the corresponding value of $c_1(z = 0, t = 0.4) = 1$ since the epithelial concentration is driven by the blood concentration through BC (3.3b).
- $c_2(r = r_{2u}, z = 0, t = 0.4) = 0.1832$ follows $c_2(r = r_{2l}, z = 0, t = 0.4) = 0.9987$, for example, as the concentration in the epithelial membrane responds to $c_1(z = 0, t) = 1$
- $c_3(r = r_{3l}, z = 0, t = 0.4) = 0.1828$ follows $c_2(r = r_{2u}, z = 0, t = 0.4) = 0.1832$, for example, as the concentration in the brain tissue responds to $c_2(r = r_{2u}, z = 0, t = 0.4)$
- $c_3(r = r_{3u}, z = 0, t = 0.4) = 0.0265$ follows $c_3(r = r_{3l}, z = 0, t = 0.4) = 0.1828$, for example, as the concentration in the outer brain tissue responds to $c_3(r = r_{3l}, z = 0, t = 0.4)$
- BC (3.4c) is not obvious from the solution in Table 3.1, but this no flux or zero slope condition is apparent in the graphical output that follows.

- The computational effort to produce the solution is acceptable, `ncall = 1114`.

The graphical output follows.

The discontinuity at $z = 0$ is appproximated by a line of finite slope due to the gridding of 21 points in z. t is plotted parametrically in Figs. 3.1 for $t = 0, 0.4, \ldots, 2$ with the bottom curve (horizontal line at zero) corresponding to the homogeneous ICs (Table 3.1, $t = 0$).

The increase in the tissue concentration, for example, $c_3(r, z, t)$ in Figs. 3.1d,e, results from transfer from the membrane. This increase is moderated by the reaction term $-k_1 c_3(r, z, t)$ in eq. (3.1c), which could represent, for example, consumption of O_2 or nutrients from the blood.

The second case programmed in Listing 3.1 corresponds to no transfer from the blood,

```
#
# Model parameters
     u10=0;        u20=0;
     u30=0;        v_c1=1;
     c1e=1;        k1=1;
```

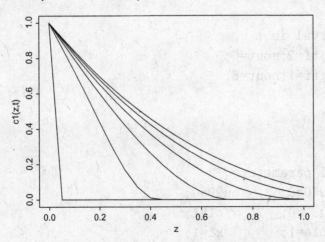

Figure 3.1a: Numerical solution $c_1(z, t)$, $k_{12f} = k_{12r} = k_{23f} = k_{23r} = 1$

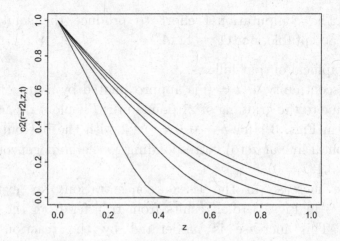

Figure 3.1b: Numerical solution $c_2(r = r_{2l}, z, t)$, $k_{12f} = k_{12r} = k_{23f} = k_{23r} = 1$

```
 D2=1.0e-06;  D3=1.0e-06;
     k12f=1;      k12r=1;
     k23f=1;      k23r=1;
 #   k12f=0;      k12r=0;
 #   k23f=0;      k23r=0;


 #
 # Interval in t
   t0=0;tf=2;nout=6;
 # t0=0;tf=1;nout=6;
```

changed to

```
 #
 # Model parameters
       u10=0;       u20=0;
       u30=0;       v_c1=1;
       c1e=1;        k1=1;
   D2=1.0e-06;  D3=1.0e-06;
```

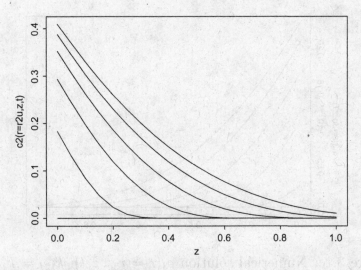

Figure 3.1c: Numerical solution $c_2(r = r_{2u}, z, t)$, $k_{12f} = k_{12r} = k_{23f} = k_{23r} = 1$

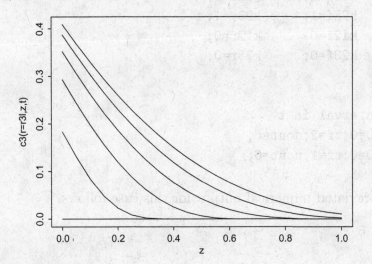

Figure 3.1d: Numerical solution $c_3(r = r_{3l}, z, t)$, $k_{12f} = k_{12r} = k_{23f} = k_{23r} = 1$

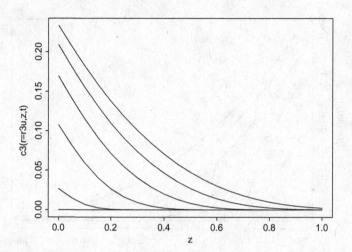

Figure 3.1e: Numerical solution $c_3(r = r_{3u}, z, t)$, $k_{12f} = k_{12r} = k_{23f} = k_{23r} = 1$

```
#     k12f=1;      k12r=1;
#     k23f=1;      k23r=1;
      k12f=0;      k12r=0;
      k23f=0;      k23r=0;


#
# Interval in t
# t0=0;tf=2;nout=6;
  t0=0;tf=1;nout=6;
```

Abbreviated numerical output for this case follows.

```
[1] 6

[1] 274

  ncall = 1354

      t      z    c1(z,t)    c21(r,z,t)   c2u(r,z,t)
```

t	z	c1(z,t)	c3l(r,z,t)	c3u(r,z,t)
0.00	0.00	1.0000	0.0000	0.0000
0.00	0.00	1.0000	0.0000	0.0000
0.00	0.10	0.0000	0.0000	0.0000
0.00	0.10	0.0000	0.0000	0.0000
0.00	0.20	0.0000	0.0000	0.0000
0.00	0.20	0.0000	0.0000	0.0000
0.00	0.30	0.0000	0.0000	0.0000
0.00	0.30	0.0000	0.0000	0.0000
0.00	0.40	0.0000	0.0000	0.0000
0.00	0.40	0.0000	0.0000	0.0000
0.00	0.50	0.0000	0.0000	0.0000
0.00	0.50	0.0000	0.0000	0.0000
0.00	0.60	0.0000	0.0000	0.0000
0.00	0.60	0.0000	0.0000	0.0000
0.00	0.70	0.0000	0.0000	0.0000
0.00	0.70	0.0000	0.0000	0.0000
0.00	0.80	0.0000	0.0000	0.0000
0.00	0.80	0.0000	0.0000	0.0000
0.00	0.90	0.0000	0.0000	0.0000
0.00	0.90	0.0000	0.0000	0.0000
0.00	1.00	0.0000	0.0000	0.0000
0.00	1.00	0.0000	0.0000	0.0000

t	z	c1(z,t)	c2l(r,z,t)	c2u(r,z,t)
t	z	c1(z,t)	c3l(r,z,t)	c3u(r,z,t)
0.20	0.00	1.0000	0.0000	0.0000
0.20	0.00	1.0000	0.0000	0.0000
0.20	0.10	0.9442	0.0000	0.0000
0.20	0.10	0.9442	0.0000	0.0000
0.20	0.20	0.5523	0.0000	0.0000
0.20	0.20	0.5523	0.0000	0.0000
0.20	0.30	0.0033	0.0000	0.0000
0.20	0.30	0.0033	0.0000	0.0000
0.20	0.40	-0.0000	0.0000	0.0000
0.20	0.40	-0.0000	0.0000	0.0000
0.20	0.50	-0.0000	0.0000	0.0000
0.20	0.50	-0.0000	0.0000	0.0000
0.20	0.60	-0.0000	0.0000	0.0000
0.20	0.60	-0.0000	0.0000	0.0000
0.20	0.70	-0.0000	0.0000	0.0000
0.20	0.70	-0.0000	0.0000	0.0000
0.20	0.80	-0.0000	0.0000	0.0000
0.20	0.80	-0.0000	0.0000	0.0000
0.20	0.90	-0.0000	0.0000	0.0000
0.20	0.90	-0.0000	0.0000	0.0000
0.20	1.00	-0.0000	0.0000	0.0000

0.20   1.00   -0.0000   0.0000   0.0000
                 .   .
                 .   .

                 .   .

Output for t=0.4,0.6,0.8 removed
                 .   .

                 .   .

t	z	c1(z,t)	c2l(r,z,t)	c2u(r,z,t)
t	z	c1(z,t)	c3l(r,z,t)	c3u(r,z,t)
1.00	0.00	1.0000	0.0000	0.0000
1.00	0.00	1.0000	0.0000	0.0000
1.00	0.10	1.0000	0.0000	0.0000
1.00	0.10	1.0000	0.0000	0.0000
1.00	0.20	1.0000	0.0000	0.0000
1.00	0.20	1.0000	0.0000	0.0000
1.00	0.30	1.0000	0.0000	0.0000
1.00	0.30	1.0000	0.0000	0.0000
1.00	0.40	1.0000	0.0000	0.0000
1.00	0.40	1.0000	0.0000	0.0000
1.00	0.50	1.0000	0.0000	0.0000
1.00	0.50	1.0000	0.0000	0.0000
1.00	0.60	0.9999	0.0000	0.0000
1.00	0.60	0.9999	0.0000	0.0000
1.00	0.70	0.9991	0.0000	0.0000
1.00	0.70	0.9991	0.0000	0.0000

1.00	0.80	0.9906	0.0000	0.0000
1.00	0.80	0.9906	0.0000	0.0000
1.00	0.90	0.8892	0.0000	0.0000
1.00	0.90	0.8892	0.0000	0.0000
1.00	1.00	0.4876	0.0000	0.0000
1.00	1.00	0.4876	0.0000	0.0000

Table 3.2: Output for eqs. (3.1) to (3.4),
$$k_{12f} = k_{12r} = k_{23f} = k_{23r} = 0$$

We can note the following details about this output.

- $c_2(r, z, t) = c_3(r, z, t) = 0$ since there is no transfer from the blood to the membrane and tissue ($k_{12f} = k_{12r} = k_{23f} = k_{23r} = 0$).
- This case of no transfer for eqs. (3.1) is the same as for the case of $k_{12f} = k_{12r} = 0$ for eqs. (2.1c,h). Therefore, the solutions should agree, which is confirmed by comparing Tables 2.2 and 3.2 (there is a small numerical difference which is apparently due to integrating $21 + (21)(6) = 147$ MOL/ODEs for eqs. (2.1c,h) and $21 + (21)(6)(2) = 273$ MOL/ODEs for eqs. (3.1)).

The agreement between the two PDE model of Chapter 2 (eqs. (2.1c,h)) and the three PDE model of Chapter 3 (eqs. (3.1)) is confirmed by a comparison of Figs. 2.2a and 3.2a. These figures indicate a small gridding effect with nz=21. To investigate this effect, the following changes were made to Listing 3.1 for nz=51.

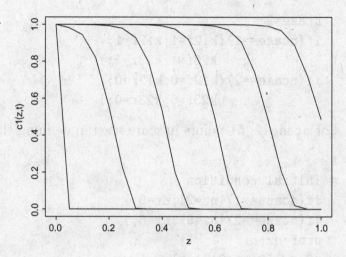

Figure 3.2a: Numerical solution $c_1(z,t)$, $k_{12f} = k_{12r} = k_{23f} = k_{23r} = 0$

- The ODE/MOL routine name is `pde1a_1`, but it is the same rotuine as `pde1a` of Listing 3.2. The name change was made to indicate the case of `nz=51` rather than `nz=21`.

```
#
# Access functions for numerical solution
  setwd("f:/BBB/chap3");
  source("pde1a_1.R");
  source("dss004.R");
  source("dss044.R");
  source("van1.R");
```

- `ncase=2` corresponds to no transfer from the blood.

```
#
# Select case
```

```
ncase=2;
if(ncase==1){k12f=1;k12r=1;
               k23f=1;k23r=1;}
if(ncase==2){k12f=0;k12r=0;
               k23f=0;k23r=0;}
```

- For ncase=2, 51 points in z are used in defining the IC.

```
#
# Initial condition
 if(ncase==1){nz=21;nr=6;}
 if(ncase==2){nz=51;nr=6;}
 nznr=nz*nr;
 u0=rep(0,nz+2*nznr);
 for(iz in 1:nz){
   u0[iz]=u10;
 for(ir in 1:nr){
   izir=(iz-1)*nr+ir;
   u0[nz+izir]     =u20;
   u0[nz+izir+nznr]=u30;
 }
 }
```

- For ncase=2, $0 \leq t \leq t_f = 1$ for better resolution of the graphical solution.

```
#
# Interval in t
 if(ncase==1){t0=0;tf=2;nout=6;}
 if(ncase==2){t0=0;tf=1;nout=6;}
 tout=seq(from=t0,to=tf,by=(tf-t0)/(nout-1));
 ncall=0;
```

- pde1a_1 is called by lsodes.

```
#
# ODE integration
```

```
out=lsodes(y=u0,times=tout,func=pde1a_1,
    sparsetype="sparseint",rtol=1e-6,
    atol=1e-6,maxord=5);
nrow(out)
ncol(out)
```

Execution of the main program in Listing 3.1 with these changes gives the following abbreviated numerical output.

[1] 6

[1] 664

ncall = 3388

t	z	c1(z,t)	c2l(r,z,t)	c2u(r,z,t)
t	z	c1(z,t)	c3l(r,z,t)	c3u(r,z,t)
0.00	0.00	1.0000	0.0000	0.0000
0.00	0.00	1.0000	0.0000	0.0000
0.00	0.04	0.0000	0.0000	0.0000
0.00	0.04	0.0000	0.0000	0.0000

Output for z=0.08 to 0.92 removed

0.00	0.96	0.0000	0.0000	0.0000
0.00	0.96	0.0000	0.0000	0.0000
0.00	1.00	0.0000	0.0000	0.0000
0.00	1.00	0.0000	0.0000	0.0000

```
      Output for t=0.2 to 0.8 removed
```

t	z	c1(z,t)	c21(r,z,t)	c2u(r,z,t)
t	z	c1(z,t)	c31(r,z,t)	c3u(r,z,t)
1.00	0.00	1.0000	0.0000	0.0000
1.00	0.00	1.0000	0.0000	0.0000
1.00	0.04	1.0000	0.0000	0.0000
1.00	0.04	1.0000	0.0000	0.0000

```
      Output for z=0.08 to 0.92 removed
```

1.00	0.96	0.7484	0.0000	0.0000
1.00	0.96	0.7484	0.0000	0.0000
1.00	1.00	0.2862	0.0000	0.0000
1.00	1.00	0.2862	0.0000	0.0000

Table 3.3: Output for eqs. (3.1) to (3.4),
$k_{12f} = k_{12r} = k_{23f} = k_{23r} = 0$, nz=51

We can note the following details about this output.

- The homogeneous (zero) ICs are confirmed at $t = 0$.
- The output is for $t = 0, 0.2, \ldots, 1$ (corresponding to tf=1).

- The output is for $z = 0, 0.04, \ldots, 1$ (corresponding to nz=51, by=2).
- out, the array with the PDE solution, has dimensions out(6,51+2*51*6+1=664) as discussd previously. The offset +1 results from the inclusion of t as the first element in each of the 6 664-vectors.
- The membrane and tissue concentrations remain at the IC, $c_2(z, r, t = 0) = c_3(z, r, t = 0) = 0$ as expected since there is no transfer from the blood ($k_{12f} = k_{12r} = k_{23f} = k_{23r} = 0$).
- The BC $c_1(z = 1, t) = 1$ constitutes a unit step that moves through z with increasing t so that at $t = 1$, this step occurs near $z = 1$. This feature of the solution is clear in Fig. 3.2a_51.
- A comparison of Figs. 3.2a and 3.2a_51 indicates that the gridding effect with nz=21 has been removed with nz=51 and the moving unit step is better represented with nz=51.

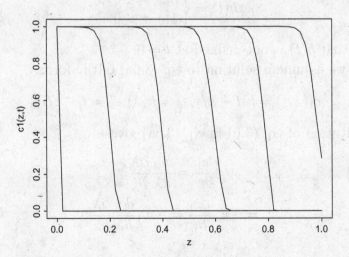

Figure 3.2a_51: Numerical solution $c_1(z, t)$, $k_{12f} = k_{12r} = k_{23f} = k_{23r} = 0$, nz=51

- The computational effort to produce the solution is increased, `ncall` = 3388, which results in a substantially longer compute time since there is also a significant increase in the number of MOL/ODEs, that is, 273 for `nz=21` increases to 663 for `nz=51`.

The traveling unit step in Fig. 3.2_51 is an exact solution to eq. (3.1a) which for no transfer ($k_{12f} = k_{12r} = k_{23f} = k_{23r} = 0$) is the linear advection equation

$$\frac{\partial c_1(z,t)}{\partial t} = -v_{c1}\frac{\partial c_1(z,t)}{\partial z} \qquad (3.5a)$$

with IC (3.2a) and BC (3.2b). For the present analysis, we take

$$c_1(z, t = 0) = c_{10} = 0 \qquad (3.5b)$$

$$c_1(z = 0, t) = c_{1e} = h(t) \qquad (3.5c)$$

where $h(t)$ is the unit step (Heaviside) function

$$h(t) = \begin{cases} 0, & t < 0 \\ 1, & t > 0 \end{cases} \qquad (3.5d)$$

Note that $h(t)$ is not defined at $t = 0$.

If we assume a solution to eq. (3.5a) of the form

$$c_1(z,t) = h(t - z/v_{c1}) = h(\lambda); \quad \lambda = t - z/v_{c1} \qquad (3.6)$$

substitution of eq. (3.6) in eq. (3.5a) gives

$$\frac{\partial c_1}{\partial t} = \frac{dc_1}{d\lambda}\frac{\partial \lambda}{\partial t}$$

$$= -v_{c1}\frac{\partial c_1}{\partial z} = -v_{c1}\frac{dc_1}{d\lambda}\frac{\partial \lambda}{\partial z}$$

Since

$$\frac{\partial \lambda}{\partial t} = 1; \quad \frac{\partial \lambda}{\partial z} = -1/v_{c1}$$

substitution in the preceding equation gives

$$\frac{\partial c_1}{\partial t} = \frac{dc_1}{d\lambda}(1)$$

$$= -v_{c1}\frac{\partial c_1}{\partial z} = -v_{c1}\frac{dc_1}{d\lambda}(-1/v_{c1}) = \frac{dc_1}{d\lambda}$$

so that eq. (3.6) is a solution to eq. (3.5a). It also satifies IC (3.5b) and BC (3.5c) so it is a solution for $c_1(z,t)$ for the special case of no transfer from the blood ($k_{12f} = k_{12r} = k_{23f} = k_{23r} = 0$).

$c_1(z,t)$ of eq. (3.6) is a unit step, traveling left to right with velocity v_{c1} for $v_{c1} > 0$. The approximation of this solution in Fig. 3.2_51 indicates no numerical oscillation and limited numerical (axial) diffusion, in contrast to the solution of eq. (3.5a) with FDs that generally produce oscillation and diffusion as numerical errors ([1]). The lack of numerical errors with vanl is the principal reason for using a flux limiter such as the van Leer limiter rather than a FD in analyzing strongly convective (hyperbolic) PDEs.

As alternative approach to eq. (3.1a) would be to use a blood velocity that varies radially across the capillary (such as Poiseuille's law if the blood is assumed to be a Newtonian fluid). This has the possible advantage of a solution that does not propagate steep moving fronts and/or discontinuities as does eq. (3.5a), but c_1 would also become 2D, that is $c_1(r,z,t)$ and gridding in r as well as z would be required.

This completes the discussion of the three PDE model of eqs. (3.1) to (3.4). The solutions are now displayed in 3D with the utility persp in place of matplot.

(3.3) 3D Visualization of the PDE Solutions

The routines for 3D visulaization are considered next, starting with the main program.

(3.3.1) Main program

The following main program is a variant of the main program in Listing 3.1.

```
#
#   Three PDE BBB model
#
# Delete previous workspaces
  rm(list=ls(all=TRUE))
#
# Access ODE integrator
  library("deSolve");
#
# Access functions for numerical solution
  setwd("f:/BBB/chap3");
  source("pde1b.R");
  source("dss004.R");
  source("dss044.R");
  source("vanl.R");
#
# Model parameters
        u10=0;          u20=0;
        u30=0;          v_c1=1;
        c1e=1;          k1=1;
  D2=1.0e-04; D3=1.0e-04;
      k12f=1;         k12r=1;
      k23f=1;         k23r=1;
#
# Initial condition
  nz=21;nr=6;
  nznr=nz*nr;
  u0=rep(0,nz+2*nznr);
  for(iz in 1:nz){
    u0[iz]=u10;
```

```
  for(ir in 1:nr){
    izir=(iz-1)*nr+ir;
    u0[nz+izir]      =u20;
    u0[nz+izir+nznr]=u30;
  }
  }
#
# Grid in z
  zl=0;zu=1;
  z=seq(from=zl,to=zu,by=(zu-zl)/(nz-1));
#
# Grid in r for c2
  r2l=1.0e-02;r2u=2.0e-02;
  r2=seq(from=r2l,to=r2u,by=(r2u-r2l)/(nr-1));
#
# Grid in r for c3
  r3l=2.0e-02;r3u=3.0e-02;
  r3=seq(from=r3l,to=r3u,by=(r3u-r3l)/(nr-1));
#
# Interval in t
  t0=0;tf=2;nout=21;
  tout=seq(from=t0,to=tf,by=(tf-t0)/(nout-1));
  ncall=0;
#
# ODE integration
  out=lsodes(y=u0,times=tout,func=pde1b,
      sparsetype="sparseint",rtol=1e-6,
      atol=1e-6,maxord=5);
  nrow(out)
  ncol(out)
#
# Store solution
  c1 =matrix(0,nrow=nz,ncol=nout);
  c2l=matrix(0,nrow=nz,ncol=nout);
```

```
  c2u=matrix(0,nrow=nz,ncol=nout);
  c3l=matrix(0,nrow=nz,ncol=nout);
  c3u=matrix(0,nrow=nz,ncol=nout);
  t=rep(0,nout);
  for(it in 1:nout){
  for(iz in 1:nz){
    c1[iz,it]=out[it,iz+1];
   c2l[iz,it]=out[it,nz+1+(iz-1)*nr+1];
   c2u[iz,it]=out[it,nz+1+(iz-1)*nr+nr];
   c3l[iz,it]=out[it,nz+1+(iz-1)*nr+1+nznr];
   c3u[iz,it]=out[it,nz+1+(iz-1)*nr+nr+nznr];
        t[it]=out[it,1];
  }
  c1[1,it]=c1e;
  }
#
# Display ncall
  cat(sprintf("\n\n ncall = %2d",ncall));
#
# Display numerical solution
  for(it in 1:nout){
    if((it-1)*(it-11)*(it-21)==0){
      cat(sprintf(
      "\n\n       t       z    c1(z,t)
        c2l(r,z,t)    c2u(r,z,t)"));
      cat(sprintf(
        "\n       t       z    c1(z,t)
        c3l(r,z,t)    c3u(r,z,t)"));
      izv=seq(from=1,to=nz,by=2);
      for(iz in izv){
        cat(sprintf(
          "\n%7.2f%7.2f%10.4f%12.4f%12.4f",
          t[it],z[iz],c1[iz,it],c2l[iz,it],
          c2u[iz,it]));
```

```
      cat(sprintf(
        "\n%7.2f%7.2f%10.4f%12.4f%12.4f\n",
        t[it],z[iz],c1[iz,it],c3l[iz,it],
        c3u[iz,it]));
    }
  }
}
#
# Plot numerical solutions
#
# c1(z,t)
  persp(
    z,t,c1,theta=45,phi=45,xlim=c(zl,zu),
    ylim=c(t0,tf),xlab="z",ylab="t",
    zlab="c1(z,t)");
#
# c2(r=r2l,z,t)
  persp(
    z,t,c2l,theta=45,phi=45,xlim=c(zl,zu),
    ylim=c(t0,tf),xlab="z",ylab="t",
    zlab="c2(r=r2l,z,t)");
#
# c2(r=r2u,z,t)
  persp(
    z,t,c2u,theta=45,phi=45,xlim=c(zl,zu),
    ylim=c(t0,tf),xlab="z",ylab="t",
    zlab="c2(r=r2u,z,t)");
#
# c3(r=r3l,z,t)
  persp(
    z,t,c3l,theta=45,phi=45,xlim=c(zl,zu),
    ylim=c(t0,tf),xlab="z",ylab="t",
    zlab="c3(r=r3l,z,t)");
#
```

```
# c3(r=r3u,z,t)
  persp(
    z,t,c3u,theta=45,phi=45,xlim=c(zl,zu),
    ylim=c(t0,tf),xlab="z",ylab="t",
    zlab="c3(r=r3u,z,t)");
```

Listing 3.3: Main program for eqs. (3.1) to (3.4), `persp`

We can note the following details about Listing 3.3.

- Previous workspaces are deleted and the ODE integrator library `deSolve` and subordinates routines are accessed.

```
#
#  Three PDE BBB model
#
# Delete previous workspaces
  rm(list=ls(all=TRUE))
#
# Access ODE integrator
  library("deSolve");
#
# Access functions for numerical solution
  setwd("f:/BBB/chap3");
  source("pde1b.R");
  source("dss004.R");
  source("dss044.R");
  source("van1.R");
```

pde1b is the ODE/MOL routine.
- The model parameters are defined numerically for the case of mass transfer (k12f=1,...,k23r=1).

```
#
# Model parameters
      u10=0;        u20=0;
```

```
   u30=0;        v_c1=1;
   c1e=1;        k1=1;
D2=1.0e-4;  D3=1.0e-04;
  k12f=1;      k12r=1;
  k23f=1;      k23r=1;
```

- ICs (3.2a), (3.3a), (3.4a) are implemented on a grid in z of 21 points and a grid in r or 6 points.

```
#
# Initial condition
  nz=21;nr=6;
  nznr=nz*nr;
  u0=rep(0,nz+2*nznr);
  for(iz in 1:nz){
    u0[iz]=u10;
    for(ir in 1:nr){
    izir=(iz-1)*nr+ir;
    u0[nz+izir]     =u20;
    u0[nz+izir+nznr]=u30;
    }
    }
```

The total number of MOL/ODEs (length of u0) is 21+2*21*6 = 273.

- The grid in z for $c_1(z,t)$, $c_2(r,z,t)$, $c_3(r,z,t)$ and the grids in r for $c_2(r,z,t)$, $c_3(r,z,t)$ are specified.

```
#
# Grid in z
  zl=0;zu=1;
  z=seq(from=zl,to=zu,by=(zu-zl)/(nz-1));
#
# Grid in r for c2
  r2l=1.0e-02;r2u=2.0e-02;
  r2=seq(from=r2l,to=r2u,by=(r2u-r2l)/(nr-1));
```

```
#
# Grid in r for c3
  r3l=2.0e-02;r3u=3.0e-02;
  r3=seq(from=r3l,to=r3u,by=(r3u-r3l)/(nr-1));
```

- The interval in t is $0 \leq t \leq t = t_f = 2$. 21 output points
  in t are specified rather 6 as in Listing 3.1 to increase the
  resolution in t in the 3D visualization.

```
#
# Interval in t
  t0=0;tf=2;nout=21;
  tout=seq(from=t0,to=tf,by=(tf-t0)/(nout-1));
  ncall=0;
```

- The 273 MOL/ODEs are integrated with lsodes. The
  MOL/ODE routine is pde1b considered subsequently.

```
#
# ODE integration
  out=lsodes(y=u0,times=tout,func=pde1b,
      sparsetype="sparseint",rtol=1e-6,
      atol=1e-6,maxord=5);
  nrow(out)
  ncol(out)
```

- The solutions are placed in arrays c1, c2l, c2u, c3l,
  c3u and t is placed in vector t. This use of the solution
  matrix out is discussed in detail after Listing 3.1.

```
#
# Store solution
  c1 =matrix(0,nrow=nz,ncol=nout);
  c2l=matrix(0,nrow=nz,ncol=nout);
  c2u=matrix(0,nrow=nz,ncol=nout);
  c3l=matrix(0,nrow=nz,ncol=nout);
  c3u=matrix(0,nrow=nz,ncol=nout);
```

```
t=rep(0,nout);
for(it in 1:nout){
for(iz in 1:nz){
  c1[iz,it]=out[it,iz+1];
  c2l[iz,it]=out[it,nz+1+(iz-1)*nr+1];
  c2u[iz,it]=out[it,nz+1+(iz-1)*nr+nr];
  c3l[iz,it]=out[it,nz+1+(iz-1)*nr+1+nznr];
  c3u[iz,it]=out[it,nz+1+(iz-1)*nr+nr+nznr];
      t[it]=out[it,1];
}
c1[1,it]=c1e;
}
```

- The number of calls to pde1b is displayed.

```
#
# Display ncall
  cat(sprintf("\n\n ncall = %2d",ncall));
```

- Selected solution values are displayed numerically for $t = 0, 1, 2$ (or it=1,11,21). Every second value of z is selected with by=2.

```
#
# Display numerical solution
  for(it in 1:nout){
    if((it-1)*(it-11)*(it-21)==0){
      cat(sprintf(
      "\n\n      t      z    c1(z,t)
      c2l(r,z,t)    c2u(r,z,t)"));
      cat(sprintf(
      "\n      t      z    c1(z,t)
      c3l(r,z,t)    c3u(r,z,t)"));
      izv=seq(from=1,to=nz,by=2);
      for(iz in izv){
        cat(sprintf(
```

```
            "\n%7.2f%7.2f%10.4f%12.4f%12.4f",
            t[it],z[iz],c1[iz,it],c2l[iz,it],
            c2u[iz,it]));
          cat(sprintf(
            "\n%7.2f%7.2f%10.4f%12.4f%12.4f\n",
            t[it],z[iz],c1[iz,it],c3l[iz,it],
            c3u[iz,it]));
      }
    }
  }
```

- $c_1(z,t)$ to $c_3(r = r_{3u}, z, t)$ are plotted with `persp` in 3D perspective, that is, against z and t.

```
#
# Plot numerical solutions
#
# c1(z,t)
  persp(
    z,t,c1,theta=45,phi=45,xlim=c(zl,zu),
    ylim=c(t0,tf),xlab="z",ylab="t",
    zlab="c1(z,t)");
#
# c2(r=r2l,z,t)
  persp(
    z,t,c2l,theta=45,phi=45,xlim=c(zl,zu),
    ylim=c(t0,tf),xlab="z",ylab="t",
    zlab="c2(r=r2l,z,t)");
#
# c2(r=r2u,z,t)
  persp(
    z,t,c2u,theta=45,phi=45,xlim=c(zl,zu),
    ylim=c(t0,tf),xlab="z",ylab="t",
    zlab="c2(r=r2u,z,t)");
#
```

```
# c3(r=r3l,z,t)
  persp(
    z,t,c3l,theta=45,phi=45,xlim=c(zl,zu),
    ylim=c(t0,tf),xlab="z",ylab="t",
    zlab="c3(r=r3l,z,t)");
#
# c3(r=r3u,z,t)
  persp(
    z,t,c3u,theta=45,phi=45,xlim=c(zl,zu),
    ylim=c(t0,tf),xlab="z",ylab="t",
    zlab="c3(r=r3u,z,t)");
```

Some additional explanation of these calls to `persp` follows.

- As an example, $c_1(z,t)$ is plotted against z, t with

  ```
  #
  # c1(z,t)
    persp(z,t,c1,...);
  ```

- The orientation of the 3D plot in the $x - y$ plane is defined with the angle `theta`. This value was determined by trial and error to give a clear perspective of the 3D plot.

  ```
  theta=45
  ```

- The orientation of the 3D plot relative to the z axis is defined with the angle `phi`. Again, this value was determined by trial and error to give a clear perspective of the 3D plot.

  ```
  phi=45
  ```

- The endpoints of the z and t axes are defined with

  ```
  xlim=c(zl,zu),ylim=c(t0,tf)
  ```

The limits of the z (c_1) axis could also be set with zlim. Here default values (automatic scaling in z) are used.

– Various labels are added to the 3D plots. for example, for $c_1(z,t)$.

```
xlab="z",ylab="t",zlab="c1(z,t)"
```

The utility persp is part of the basic R system and does not have to be accessed specifically.

The MOL/ODE routine pde1b is considered next.

(3.3.2) ODE/MOL routine

Function pde1b is the same as pde1a in Listing 3.2. The change in name was made to give a separate, complete set of routines for the 3D visualization.

(3.3.3) Model output

The numerical and graphical output from the main program of Listing 3.3 and pde1b is discussed next.

```
[1] 21

[1] 274

ncall = 1046
```

t	z	c1(z,t)	c2l(r,z,t)	c2u(r,z,t)
t	z	c1(z,t)	c3l(r,z,t)	c3u(r,z,t)
0.00	0.00	1.0000	0.0000	0.0000
0.00	0.00	1.0000	0.0000	0.0000
0.00	0.10	0.0000	0.0000	0.0000
0.00	0.10	0.0000	0.0000	0.0000

0.00	0.20	0.0000	0.0000	0.0000
0.00	0.20	0.0000	0.0000	0.0000
0.00	0.30	0.0000	0.0000	0.0000
0.00	0.30	0.0000	0.0000	0.0000
0.00	0.40	0.0000	0.0000	0.0000
0.00	0.40	0.0000	0.0000	0.0000
0.00	0.50	0.0000	0.0000	0.0000
0.00	0.50	0.0000	0.0000	0.0000
0.00	0.60	0.0000	0.0000	0.0000
0.00	0.60	0.0000	0.0000	0.0000
0.00	0.70	0.0000	0.0000	0.0000
0.00	0.70	0.0000	0.0000	0.0000
0.00	0.80	0.0000	0.0000	0.0000
0.00	0.80	0.0000	0.0000	0.0000
0.00	0.90	0.0000	0.0000	0.0000
0.00	0.90	0.0000	0.0000	0.0000
0.00	1.00	0.0000	0.0000	0.0000
0.00	1.00	0.0000	0.0000	0.0000
t	z	$c1(z,t)$	$c2l(r,z,t)$	$c2u(r,z,t)$
t	z	$c1(z,t)$	$c3l(r,z,t)$	$c3u(r,z,t)$
1.00	0.00	1.0000	0.9990	0.3264
1.00	0.00	1.0000	0.3259	0.1416
1.00	0.10	0.8088	0.8079	0.2342

t	z	c1(z,t)	c21(r,z,t)	c2u(r,z,t)
1.00	0.10	0.8088	0.2338	0.0884
1.00	0.20	0.6319	0.6312	0.1566
1.00	0.20	0.6319	0.1563	0.0488
1.00	0.30	0.4716	0.4709	0.0956
1.00	0.30	0.4716	0.0954	0.0231
1.00	0.40	0.3289	0.3284	0.0512
1.00	0.40	0.3289	0.0511	0.0088
1.00	0.50	0.2068	0.2064	0.0224
1.00	0.50	0.2068	0.0223	0.0023
1.00	0.60	0.1099	0.1096	0.0070
1.00	0.60	0.1099	0.0070	0.0004
1.00	0.70	0.0432	0.0431	0.0011
1.00	0.70	0.0432	0.0011	0.0000
1.00	0.80	0.0072	0.0071	0.0000
1.00	0.80	0.0072	0.0000	0.0000
1.00	0.90	0.0000	0.0000	-0.0000
1.00	0.90	0.0000	-0.0000	-0.0000
1.00	1.00	-0.0000	-0.0000	-0.0000
1.00	1.00	-0.0000	-0.0000	-0.0000
t	z	c1(z,t)	c21(r,z,t)	c2u(r,z,t)
t	z	c1(z,t)	c31(r,z,t)	c3u(r,z,t)
2.00	0.00	1.0000	0.9991	0.4090
2.00	0.00	1.0000	0.4086	0.2333

2.00	0.10	0.8405	0.8397	0.3300
2.00	0.10	0.8405	0.3296	0.1819
2.00	0.20	0.6969	0.6962	0.2601
2.00	0.20	0.6969	0.2599	0.1374
2.00	0.30	0.5701	0.5695	0.2005
2.00	0.30	0.5701	0.2002	0.1004
2.00	0.40	0.4588	0.4583	0.1504
2.00	0.40	0.4588	0.1502	0.0706
2.00	0.50	0.3620	0.3615	0.1093
2.00	0.50	0.3620	0.1092	0.0474
2.00	0.60	0.2786	0.2783	0.0765
2.00	0.60	0.2786	0.0764	0.0301
2.00	0.70	0.2080	0.2077	0.0510
2.00	0.70	0.2080	0.0510	0.0179
2.00	0.80	0.1495	0.1493	0.0321
2.00	0.80	0.1495	0.0321	0.0097
2.00	0.90	0.1024	0.1022	0.0188
2.00	0.90	0.1024	0.0188	0.0048
2.00	1.00	0.0667	0.0666	0.0103
2.00	1.00	0.0667	0.0103	0.0022

Table 3.4: Output for eqs. (3.1) to (3.4),
$k_{12f} = k_{12r} = k_{23f} = k_{23r} = 1$, persp

We can note the following details about this output.

- The solution matrix out has 21 values of t for 273 MOL/ODEs.

  [1] 21

  [1] 274

  Each 273-vector also includes the value of t, that is, of length $273 + 1 = 274$.
- The homogeneous ICs for c_1, c_2, c_3 and the entering $c_1 = c_{1e} = 1$ concentration are confirmed (at $t = 0$).
- The movement through z of $c_1 = c_{1e} = 1$ at $t = 1, 2$ is indicated. This movement is also reflected in Figs. 3.3.
- The computational effort for the 273 ODEs is acceptable, ncall = 1046.

The graphical output is in Figs. 3.3.

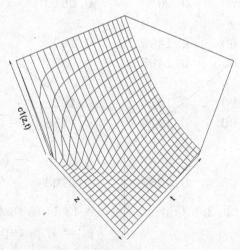

Figure 3.3a: Numerical solution $c_1(z, t)$, $k_{12f} = k_{12r} = k_{23f} = k_{23r} = 1$, persp

The propagation of the moving front through the blood is demonstrated in Fig. 3.3a.

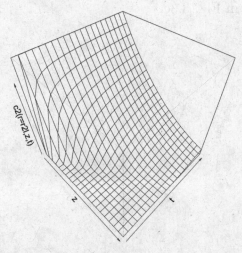

Figure 3.3b: Numerical solution $c_2(r = r_{2l}, z, t)$, $k_{12f} = k_{12r} = k_{23f} = k_{23r} = 1$, `persp`

The increase in the inner membrane concentration, $c_2(r = r_{2l}, z, t)$, in response to the increasing blood concentration is demonstrated in Fig. 3.3b.

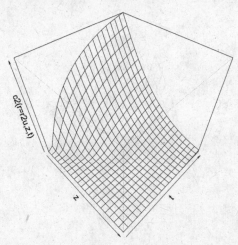

Figure 3.3c: Numerical solution $c_2(r = r_{2u}, z, t)$, $k_{12f} = k_{12r} = k_{23f} = k_{23r} = 1$, `persp`

The increase in the outer membrane concentration, $c_2(r = r_{2u}, z, t)$, in response to the increasing blood concentration is demonstrated in Fig. 3.3c.

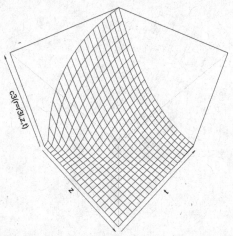

Figure 3.3d: Numerical solution $c_3(r = r_{3l}, z, t)$, $k_{12f} = k_{12r} = k_{23f} = k_{23r} = 1$, `persp`

The increase in the inner tissue concentration, $c_3(r = r_{3l}, z, t)$, in response to the increasing membrane concentration is demonstrated in Fig. 3.3d.

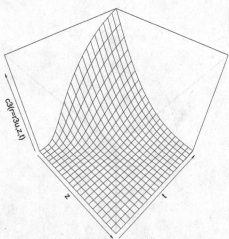

Figure 3.3e: Numerical solution $c_3(r = r_{3u}, z, t)$, $k_{12f} = k_{12r} = k_{23f} = k_{23r} = 1$, `persp`

The increase in the outer tissue concentration, $c_3(r = r_{3u}, z, t)$, in response to the increasing membrane concentration is demonstrated in Fig. 3.3e.

This completes the examples of three PDE models.

(3.4) Summary and Conclusions

A three PDE model for transfer through the blood brain barrier, eqs. (3.1) to (3.4), has been implemented and the computed solutions presented numerically, and in 2D and 3D perspective.

A small variant of eq. (3.1a) is the use of a Fick's first law transfer rate between the blood and membrane.

$$\frac{\partial c_1(z, t)}{\partial t} = -v_{c1}\frac{\partial c_1(z, t)}{\partial z} + \frac{2}{r_1}D_2\frac{\partial c_2(r = r_{2l}, z, t)}{\partial r} \qquad (3.1d)$$

The transfer term follows directly from BC (3.3b) and the solution is the same as in Table 3.1 and Figs. 3.1.

Eqs. (3.1) to (3.4) are linear in the sense that $c_1(z, t), c_2(r, z, t), c_3(r, z, t)$ and their derivatives are to the first power (degree). In the next chapter, some nonlinear forms of these equations are considered.

Reference

[1] Griffiths, G.W., and W.E. Schiesser (2012), *Traveling Wave Analysis of Partial Diffenetial Equations*, Elsevier, Burlington, MA, USA

Chapter 4

Nonlinear PDE Models

(4.1) Introduction

The preceding blood brain barrier (BBB) models were linear in the sense that the dependent variables $c_1(z, t), c_2(r, z, t)$, $c_3(r, z, t)$ and their derivatives were to the first power (degree). In this chapter we consider some nonlinear variants that include additional effects which might enhance the description of BBB dynamics.

(4.2) First Nonlinear PDE Model

The first nonlinear model has a reaction term in the $c_3(r, z, t)$ mass balance, eq. (3.1c), that is not first order.

$$\frac{\partial c_3(r, z, t)}{\partial t} = D_3 \left(\frac{\partial^2 c_3(r, z, t)}{\partial r^2} + \frac{1}{r} \frac{\partial c_3(r, z, t)}{\partial r} \right) - k_1 c_3(r, z, t)^n \tag{4.1}$$

Eq. (3.1c) has a first order reaction term, $-k_1 c_3(r, z, t)$, and eq. (4.1) has an nth order reaction term, $-k_1 c_3(r, z, t)^n$. This may seem like a trivial difference, but there are two important points.

- The reaction term is nonlinear for $n \neq 1$. While this nonlinear term is easily programmed in the ODE/MOL routine pde1a discussed subsequently, it would probably preclude an analytical solution to eqs. (3.1a,b) and (4.1).

103

- The effect of the reaction order n can be studied numerically merely by changing **n** (in the main program that follows). As might be expected, this has a small effect on the blood and membrane concentrations, $c_1(z,t), c_2(r,z,t)$, but has a significant effect on the tissue concentration $c_3(r,z,t)$.

These features of the numerical solution are demonstrated by the examples that follow in which $n = 1, 2$ (two cases).

(4.2.1) Main program

The main program, which closely parallels the main program in Listing 3.1, follows.

```
#
#  Three PDE BBB model
#
# Delete previous workspaces
  rm(list=ls(all=TRUE))
#
# Access ODE integrator
  library("deSolve");
#
# Access functions for numerical solution
  setwd("f:/BBB/chap4");
  source("pde1a.R");
  source("dss004.R");
  source("dss044.R");
  source("vanl.R");
#
# Model parameters
      u10=0;        u20=0;
      u30=0;        v_c1=1;
    c1e=10;         k1=1;
```

```
  D2=1.0e-06; D3=1.0e-06;
     k12f=1;        k12r=1;
     k23f=1;        k23r=1;
#
# Reaction order
  ncase=1;
  if(ncase==1){n=1;}
  if(ncase==2){n=2;}
#
# Initial condition
  nz=21;nr=6;
  nznr=nz*nr;
  u0=rep(0,nz+2*nznr);
  for(iz in 1:nz){
    u0[iz]=u10;
  for(ir in 1:nr){
    izir=(iz-1)*nr+ir;
    u0[nz+izir]       =u20;
    u0[nz+izir+nznr]=u30;
  }
  }
#
# Grid in z
  zl=0;zu=1;
  z=seq(from=zl,to=zu,by=(zu-zl)/(nz-1));
#
# Grid in r for c2
  r2l=1.0e-03;r2u=2.0e-03;
  r2=seq(from=r2l,to=r2u,by=(r2u-r2l)/(nr-1));
#
# Grid in r for c3
  r3l=2.0e-03;r3u=3.0e-03;
  r3=seq(from=r3l,to=r3u,by=(r3u-r3l)/(nr-1));
#
```

```
# Interval in t
  t0=0;tf=2;nout=6;
  tout=seq(from=t0,to=tf,by=(tf-t0)/(nout-1));
  ncall=0;
#
# ODE integration
  out=lsodes(y=u0,times=tout,func=pde1a,
      sparsetype="sparseint",rtol=1e-6,
      atol=1e-6,maxord=5);
  nrow(out)
  ncol(out)
#
# Store solution
  c1 =matrix(0,nrow=nz,ncol=nout);
  c2l=matrix(0,nrow=nz,ncol=nout);
  c2u=matrix(0,nrow=nz,ncol=nout);
  c3l=matrix(0,nrow=nz,ncol=nout);
  c3u=matrix(0,nrow=nz,ncol=nout);
  t=rep(0,nout);
 \for(it in 1:nout){
  for(iz in 1:nz){
     c1[iz,it]=out[it,iz+1];
    c2l[iz,it]=out[it,nz+1+(iz-1)*nr+1];
    c2u[iz,it]=out[it,nz+1+(iz-1)*nr+nr];
    c3l[iz,it]=out[it,nz+1+(iz-1)*nr+1+nznr];
    c3u[iz,it]=out[it,nz+1+(iz-1)*nr+nr+nznr];
         t[it]=out[it,1];
  }
  c1[1,it]=c1e;
  }
#
# Display ncall
  cat(sprintf("\n\n ncall = %2d",ncall));
#
```

```
# Display numerical solution
  for(it in 1:nout){
  cat(sprintf(
    "\n\n       t       z     c1(z,t)       c2l(r,z,t)
    c2u(r,z,t)"));
  cat(sprintf(
      "\n       t       z     c1(z,t)       c3l(r,z,t)
      c3u(r,z,t)"));
  izv=seq(from=1,to=nz,by=2);
  for(iz in izv){
    cat(sprintf(
      "\n%7.2f%7.2f%10.4f%12.4f%12.4f",
      t[it],z[iz],c1[iz,it],c2l[iz,it],
      c2u[iz,it]));
    cat(sprintf(
      "\n%7.2f%7.2f%10.4f%12.4f%12.4f\n",
      t[it],z[iz],c1[iz,it],c3l[iz,it],
      c3u[iz,it]));
  }
  }
#
# Plot numerical solutions
#
# c1(z,t)
  par(mfrow=c(1,1));
  matplot(
    x=z,y=c1,type="l",xlab="z",
    ylab="c1(z,t)",xlim=c(zl,zu),
    lty=1,main="",lwd=2,col="black");
#
# c2(r=r2l,z,t)
  par(mfrow=c(1,1));
  matplot(
    x=z,y=c2l,type="l",xlab="z",
```

```
    ylab="c2(r=r2l,z,t)",xlim=c(zl,zu),
    lty=1,main="",lwd=2,col="black");
#
# c2(r=r2u,z,t)
  par(mfrow=c(1,1));
  matplot(
    x=z,y=c2u,type="l",xlab="z",
    ylab="c2(r=r2u,z,t)",xlim=c(zl,zu),
    lty=1,main="",lwd=2,col="black");
#
# c3(r=rl3,z,t)
  par(mfrow=c(1,1));
  matplot(
    x=z,y=c3l,type="l",xlab="z",
    ylab="c3(r=r3l,z,t)",xlim=c(zl,zu),
    lty=1,main="",lwd=2,col="black");
#
# c3(r=ru3,z,t)
  par(mfrow=c(1,1));
  matplot(x=z,y=c3u,type="l",xlab="z",
    ylab="c3(r=r3u,z,t)",xlim=c(zl,zu),
    lty=1,main="",lwd=2,col="black");
```

Listing 4.1: Main program for eqs. (3.1a,b), (4.1), n=1,2

We can note the following details about Listing 4.1.

- Previous workspaces are deleted.

```
    #
    # Three PDE BBB model
    #
    # Delete previous workspaces
      rm(list=ls(all=TRUE))
```

- The R ODE integrator library deSolve is accessed. Then the directory with the files for the solution of eqs. (3.1a,b), (4.1) is designated. Note that setwd (set working directory) uses / rather than the usual \.

```
#
# Access ODE integrator
  library("deSolve");
#
# Access functions for numerical solution
  setwd("f:/BBB/chap4");
  source("pde1a.R");
  source("dss004.R");
  source("dss044.R");
  source("van1.R");
```

pde1a.R is the routine for the method of lines (MOL) approximation of PDEs (3.1a,b), (4.1) (discussed subsequently). dss004, dss044 (Differentiation in Space Subroutine) are library routines for calculating first and second derivatives in r by finite differences (FDs). van1 is a library routine for calculating first derivatives in z. The coding and use of these routines is discussed subsequently.

- The model parameters are defined numerically.

```
#
# Model parameters
      u10=0;        u20=0;
      u30=0;        v_c1=1;
      c1e=10;        k1=1;
  D2=1.0e-06; D3=1.0e-06;
      k12f=1;       k12r=1;
      k23f=1;       k23r=1;
```

The entering concentration, $c_1(z = 0, t) = c_{1e} = 10$ has an increased value (from Listing 3.1) to amplify the effect of the reaction term in eq. (4.1) (and thereby demonstrate the effect of changes in the reaction order n).

- Two cases are programmed for a first order ncase=1) and a second order (ncase=2) reaction.

```
#
# Reaction order
  ncase=1;
  if(ncase==1){n=1;}
  if(ncase==2){n=2;}
```

- The initial conditions (ICs) for eqs. (3.1a,b), (4.1) are placed in a vector u0 of length nz + 2*nz*nr = 21 + 2*21*6 = 273.

```
#
# Initial condition
  nz=21;nr=6;
  nznr=nz*nr;
  u0=rep(0,nz+2*nznr);
  for(iz in 1:nz){
    u0[iz]=u10;
  for(ir in 1:nr){
    izir=(iz-1)*nr+ir;
    u0[nz+izir]      =u20;
    u0[nz+izir+nznr]=u30;
  }
  }
```

- A spatial grid of nz=21 points is defined for $z_l = 0 \le z \le z_u = 1$, so that $z = 0, 0.05, \ldots, 1$.

```
#
# Grid in z
```

```
zl=0;zu=1;
z=seq(from=zl,to=zu,by=(zu-zl)/(nz-1));
```

- A spatial grid of 6 points is defined for the membrane, $r_{2l} = 0.001 \leq r_2 \leq r_{2u} = 0.002$, so that $r_2 = 0.001$, $0.0012, \ldots, 0.002$.

```
#
# Grid in r for c2
  r2l=1.0e-03;r2u=2.0e-03;
  r2=seq(from=r2l,to=r2u,by=(r2u-r2l)/(nr-1));
```

The small dimensions `r2l`, `r2u` reflect the thin membrane.

- A spatial grid of 6 points is defined for the tissue, $r_{3l} = 0.002 \leq r_3 \leq r_{3u} = 0.003$, so that $r_3 = 0.002$, $0.0022, \ldots, 0.003$.

```
#
# Grid in r for c3
  r3l=2.0e-03;r3u=3.0e-03;
  r3=seq(from=r3l,to=r3u,by=(r3u-r3l)/(nr-1));
```

- An interval in t of 6 points is defined for $0 \leq t \leq 2$ so that `tout=0,0.4,...,2`.

```
#
# Interval in t
  t0=0;tf=2;nout=6;
  tout=seq(from=t0,to=tf,by=(tf-t0)/(nout-1));
  ncall=0;
```

The counter for the calls to the ODE/MOL routine `pde1a` is also initialized.

- The system of 273 MOL/ODEs is integrated by the library integrator `lsodes` (available in `deSolve`). As expected, the inputs to `lsodes` are the ODE function, pde1a, the IC vector u0, and the vector of output values of t, tout. The length of u0 (e.g., 273) informs lsodes how many ODEs are to be integrated. func,y,times are reserved names.

```
#
# ODE integration
  out=lsodes(y=u0,times=tout,func=pde1a,
      sparsetype="sparseint",rtol=1e-6,
      atol=1e-6,maxord=5);
  nrow(out)
  ncol(out)
```

The numerical solution to the ODEs is returned in matrix out. In this case, out has the dimensions $nout \times (nz + 2(nz)(nr) + 1) = 6 \times 21 + (2)(21)(6) + 1 = 274$, which are confirmed by the output from nrow(out),ncol(out) (included in the numerical output considered subsequently).

The offset $273 + 1 = 274$ is required since the first element of each column has the output t (also in tout), and the $2, \ldots, 273 + 1 = 2, \ldots, 274$ column elements have the 273 ODE solutions.

- Selected solutions of the 273 ODEs returned in out by lsodes are placed in arrays c1, c2l, c2u, c3l, c3u.

```
#
# Store solution
  c1 =matrix(0,nrow=nz,ncol=nout);
  c2l=matrix(0,nrow=nz,ncol=nout);
  c2u=matrix(0,nrow=nz,ncol=nout);
  c3l=matrix(0,nrow=nz,ncol=nout);
  c3u=matrix(0,nrow=nz,ncol=nout);
  t=rep(0,nout);
  for(it in 1:nout){
```

```
for(iz in 1:nz){
    c1[iz,it]=out[it,iz+1];
   c2l[iz,it]=out[it,nz+1+(iz-1)*nr+1];
   c2u[iz,it]=out[it,nz+1+(iz-1)*nr+nr];
   c3l[iz,it]=out[it,nz+1+(iz-1)*nr+1+nznr];
   c3u[iz,it]=out[it,nz+1+(iz-1)*nr+nr+nznr];
       t[it]=out[it,1];
}
c1[1,it]=c1e;
}
```

Again, the offset +1 is required since the first element of each column of out has the value of t. $c_1(z = 0, t) =$ c1[1,it]=c1e=10 sets BC (3.2b) since this value is not returned by lsodes (it is defined algebraically in pde1a and not by the integration of an ODE at $z = 0$). $c_2(r = r_{2l}, z, t), c_2(r = r_{2u}, z, t)$ are placed in c2l, c2u. $c_3(r = r_{3l}, z, t), c_3(r = r_{3u}, z, t)$ are placed in c3l, c3u.

- The number of calls to pde1a is displayed at the end of the solution.

```
#
# Display ncall
   cat(sprintf("\n\n ncall = %2d",ncall));
```

- $t, z, c_1(z,t), c_2(r = r_{2l}, z, t), c_2(r = r_{2u}, z, t), c_3(r = r_{3l}, z, t), c_3(r = r_{3u}, z, t)$ are displayed for every second value of z from by=2.

```
#
# Display numerical solution
   for(it in 1:nout){
   cat(sprintf(
     "\n\n        t        z    c1(z,t)     c2l(r,z,t)
      c2u(r,z,t)"));
   cat(sprintf(
```

```
              "\n      t       z    c1(z,t)      c3l(r,z,t)
              c3u(r,z,t)"));
      izv=seq(from=1,to=nz,by=2);
      for(iz in izv){
        cat(sprintf(
          "\n%7.2f%7.2f%10.4f%12.4f%12.4f",
          t[it],z[iz],c1[iz,it],c2l[iz,it],
          c2u[iz,it]));
        cat(sprintf(
          "\n%7.2f%7.2f%10.4f%12.4f%12.4f\n",
          t[it],z[iz],c1[iz,it],c3l[iz,it],
          c3u[iz,it]));
      }
      }
```

- Five selected values of the solution are plotted against z
  and parametrically in t.

```
  #
  # Plot numerical solutions
  #
  # c1(z,t)
    par(mfrow=c(1,1));
    matplot(
      x=z,y=c1,type="l",xlab="z",
      ylab="c1(z,t)",xlim=c(zl,zu),
      lty=1,main="",lwd=2,col="black");
  #
  # c2(r=r2l,z,t)
    par(mfrow=c(1,1));
    matplot(
      x=z,y=c2l,type="l",xlab="z",
      ylab="c2(r=r2l,z,t)",xlim=c(zl,zu),
      lty=1,main="",lwd=2,col="black");
  #
```

```
# c2(r=r2u,z,t)
  par(mfrow=c(1,1));
  matplot(
    x=z,y=c2u,type="l",xlab="z",
    ylab="c2(r=r2u,z,t)",xlim=c(zl,zu),
    lty=1,main="",lwd=2,col="black");
#
# c3(r=rl3,z,t)
  par(mfrow=c(1,1));
  matplot(
    x=z,y=c3l,type="l",xlab="z",
    ylab="c3(r=r3l,z,t)",xlim=c(zl,zu),
    lty=1,main="",lwd=2,col="black");
#
# c3(r=ru3,z,t)
  par(mfrow=c(1,1));
  matplot(x=z,y=c3u,type="l",xlab="z",
    ylab="c3(r=r3u,z,t)",xlim=c(zl,zu),
    lty=1,main="",lwd=2,col="black");
```

The ODE/MOL routine pde1a is considered next.

(4.2.2) ODE/MOL routine

pde1a follows.

```
  pde1a=function(t,u,parms){
#
# Function pde1a computes the t derivative
# vector of c1t(z,t), c2t(r,z,t), c3t(r,z,t)
#
# One vector to one vector, two matrices
  c1=rep(0,nz);
  c2=matrix(0,nrow=nz,ncol=nr);
  c3=matrix(0,nrow=nz,ncol=nr);
```

```
    for(iz in 1:nz){
      c1[iz]=u[iz];
    for(ir in 1:nr){
      izir=(iz-1)*nr+ir;
      c2[iz,ir]=u[izir+nz];
      c3[iz,ir]=u[izir+nz+nznr];
    }
    }
#
# Boundary condition
    c1[1]=c1e;
#
# c1z
    c1z=vanl(zl,zu,nz,c1,v_c1);
#
# c2r
    c2r=matrix(0,nrow=nz,ncol=nr);
    for(iz in 1:nz){
      c2r[iz,]=dss004(r2l,r2u,nr,c2[iz,]);
    }
#
# c3r
    c3r=matrix(0,nrow=nz,ncol=nr);
    for(iz in 1:nz){
      c3r[iz,]=dss004(r3l,r3u,nr,c3[iz,]);
    }
#
# c2 BCs
    for(iz in 1:nz){
      c2r[iz,1]=
        -(1/D2)*(k12f*c1[iz]   -k12r*c2[iz,1]);
      c2r[iz,nr]=
        -(1/D2)*(k23f*c2[iz,nr]-k23r*c3[iz,1]);
    }
```

```
#
# c3 BCs
  for(iz in 1:nz){
    c3r[iz,1]=
      -(1/D3)*(k23f*c2[iz,nr]-k23r*c3[iz,1]);
    c3r[iz,nr]=0;
  }
#
# c2rr
  nl=2;nu=2;
  c2rr=matrix(0,nrow=nz,ncol=nr);
  for(iz in 1:nz){
    c2rr[iz,]=
      dss044(r2l,r2u,nr,c2[iz,],c2r[iz,],nl,nu);
  }
#
# c3rr
  nl=2;nu=2;
  c3rr=matrix(0,nrow=nz,ncol=nr);
  for(iz in 1:nz){
    c3rr[iz,]=
      dss044(r3l,r3u,nr,c3[iz,],c3r[iz,],nl,nu);
  }
#
# c1t, c2t, c3t
  c1t=rep(0,nz);
  c2t=matrix(0,nrow=nz,ncol=nr);
  c3t=matrix(0,nrow=nz,ncol=nr);
  for(iz in 1:nz){
    c1t[iz]=
      -v_c1*c1z[iz]-(2/r2l)*
      (k12f*c1[iz]-k12r*c2[iz,1]);
  for(ir in 1:nr){
    c2t[iz,ir]=D2*(c2rr[iz,ir]+
```

```
        (1/r2[ir])*c2r[iz,ir]);
     c3t[iz,ir]=D3*(c3rr[iz,ir]+
        (1/r3[ir])*c3r[iz,ir])-
        k1*c3[iz,ir]^n;
  }
  }
  c1t[1]=0;
#
# One vector, two matrices to one vector
  ut=rep(0,nz+2*nznr);
  for(iz in 1:nz){
    ut[iz]=c1t[iz];
  for(ir in 1:nr){
    izir=(iz-1)*nr+ir;
    ut[nz+izir]     =c2t[iz,ir];
    ut[nz+izir+nznr]=c3t[iz,ir];
  }
  }
#
# Increment calls to pde1a
  ncall<<-ncall+1;
#
# Return derivative vector
  return(list(c(ut)));
}
```

Listing 4.2: ODE/MOL routine for eqs. (3.1a,b), (4.1)

Listing 4.2 is similar to Listing 3.2 with the use of

```
  -k1*c3[iz,ir]^n;
```

(from Listing 4.2) in place of

```
  -k1*c3[iz,ir];
```

(from Listing 3.2) in the programming of eq. (4.1)

```
c3t[iz,ir]=D3*(c3rr[iz,ir]+
  (1/r3[ir])*c3r[iz,ir])-
  k1*c3[iz,ir]^n;
```

The output of the main program of Listing 4.1 and ODE/MOL routine of Listing 4.2 follows.

(4.2.3) Model output

Abbreviated numerical output is in Table 4.1 for **ncase=1, n=1**.

[1] 6

[1] 274

```
ncall = 1906
```

t	z	c1(z,t)	c2l(r,z,t)	c2u(r,z,t)
t	z	c1(z,t)	c3l(r,z,t)	c3u(r,z,t)
0.00	0.00	10.0000	0.0000	0.0000
0.00	0.00	10.0000	0.0000	0.0000
0.00	0.10	0.0000	0.0000	0.0000
0.00	0.10	0.0000	0.0000	0.0000
0.00	0.20	0.0000	0.0000	0.0000
0.00	0.20	0.0000	0.0000	0.0000
0.00	0.30	0.0000	0.0000	0.0000
0.00	0.30	0.0000	0.0000	0.0000
0.00	0.40	0.0000	0.0000	0.0000
0.00	0.40	0.0000	0.0000	0.0000

$\bullet$

0.00	0.50	0.0000	0.0000	0.0000
0.00	0.50	0.0000	0.0000	0.0000
0.00	0.60	0.0000	0.0000	0.0000
0.00	0.60	0.0000	0.0000	0.0000
0.00	0.70	0.0000	0.0000	0.0000
0.00	0.70	0.0000	0.0000	0.0000
0.00	0.80	0.0000	0.0000	0.0000
0.00	0.80	0.0000	0.0000	0.0000
0.00	0.90	0.0000	0.0000	0.0000
0.00	0.90	0.0000	0.0000	0.0000
0.00	1.00	0.0000	0.0000	0.0000
0.00	1.00	0.0000	0.0000	0.0000

.
.
.

Output for t = 0.40,...,1.6 removed

.
.
.

t	z	c1(z,t)	c2l(r,z,t)	c2u(r,z,t)
t	z	c1(z,t)	c3l(r,z,t)	c3u(r,z,t)
2.00	0.00	10.0000	9.9914	4.0901
2.00	0.00	10.0000	4.0860	2.3333
2.00	0.10	8.4051	8.3975	3.2996
2.00	0.10	8.4051	3.2961	1.8191
2.00	0.20	6.9690	6.9625	2.6016
2.00	0.20	6.9690	2.5987	1.3737

2.00	0.30	5.7010	5.6953	2.0042
2.00	0.30	5.7010	2.0019	1.0033
2.00	0.40	4.5880	4.5831	1.5036
2.00	0.40	4.5880	1.5017	0.7055
2.00	0.50	3.6190	3.6149	1.0926
2.00	0.50	3.6190	1.0911	0.4737
2.00	0.60	2.7852	2.7818	0.7641
2.00	0.60	2.7852	0.7630	0.3008
2.00	0.70	2.0792	2.0765	0.5099
2.00	0.70	2.0792	0.5091	0.1785
2.00	0.80	1.4940	1.4918	0.3210
2.00	0.80	1.4940	0.3204	0.0973
2.00	0.90	1.0229	1.0213	0.1879
2.00	0.90	1.0229	0.1875	0.0477
2.00	1.00	0.6666	0.6654	0.1031
2.00	1.00	0.6666	0.1029	0.0216

Table 4.1: Solution to eqs. (3.1a,b), (4.1), n=1, c1e=10

We can note the following details about this output.

- The dimensions of the solution matrix out are out(6,274), that is, 6 output points for the solution of 273 ODEs. The offset $273 + 1 = 274$ reflects the value of t at the beginning of each of the 6 273-vectors.

  [1] 6

  [1] 274

- The homogeneous (zero) ICs are confirmed at $t = 0$. This is an important check since if the ICs are not correct, the subsequent solution for $t > 0$ will not be correct. Also, the BC $c_1(z = 0, t) = c_{1e} = 10$ is confirmed,

t	z	c1(z,t)	c2l(r,z,t)	c2u(r,z,t)
t	z	c1(z,t)	c3l(r,z,t)	c3u(r,z,t)
0.00	0.00	10.0000	0.0000	0.0000
0.00	0.00	10.0000	0.0000	0.0000

- The output is for $t = 0, 0.4, \ldots, 2$ as programmed in Listing 4.1.
- The output is for $z = 0, 0.1, \ldots, 1$ as programmed in Listing 4.1.
- $c_1(z = 0, t)$ is constant according to the inlet concentration c1[1]=c1e=10. Thus, at $t = 0$, $c_1(z = 0, t = 0) = 10$ and at $t = 2$, $c_1(z = 0, t = 2) = 10$. The solution is a response to this inlet concentration.
- The drop in concentration across the membrane is significant, e.g., $c_2(r = r_{2l}, z = 0.5, t = 2) = 3.6149$, $c_2(r = r_{2u}, z = 0.5, t = 2) = 1.0926$.

t	z	c1(z,t)	c2l(r,z,t)	c2u(r,z,t)
t	z	c1(z,t)	c3l(r,z,t)	c3u(r,z,t)
2.00	0.50	3.6190	3.6149	1.0926
2.00	0.50	3.6190	1.0911	0.4737

The drop in concentration across the blood-membrane interface is small, $c_1(z = 0.5, t = 2) = 3.6190$, $c_2(r = r_{2l}, z = 0.5, t = 2) = 3.6149$. This concentration drop could be increased by decreasing the mass transfer coefficients k_{12f}, k_{12r}.

- Similarly, the drop in concentration across the membrane-tissue interface is small, $c_2(r = r_{2u}, z = 0.5, t = 2) = 1.0926$, $c_3(r = r_{3l}, z = 0.5, t = 2) = 1.0911$. This concentration drop could be increased by decreasing the mass transfer coefficients k_{23f}, k_{23r}.

- The drop in concentration across the tissue is significant, e.g., $c_3(r = r_{3l}, z = 0.5, t = 2) = 1.0911$, $c_3(r = r_{3u}, z = 0.5, t = 2) = 0.4737$.

- The no flux or zero slope condition BC $\dfrac{\partial c_3(r = r_{3u}, z, t)}{\partial r} = 0$ is not obvious from the solution in Table 4.1, but this is apparent in the graphical output that follows.

- The computational effort to produce the solution is acceptable, `ncall = 1906`.

Abbreviated graphical output is given in Figs. 4.1a,d,e.

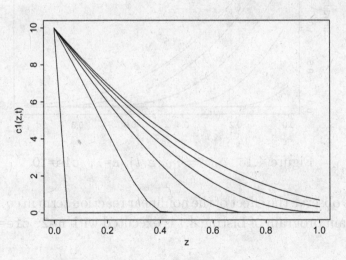

Figure 4.1a: $c_1(z, t)$, n=1, `c1e=10`

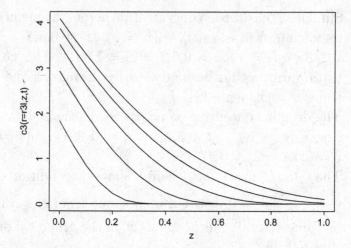

Figure 4.1d: $c_3(r = r_{3l}, z, t)$, n=1, c1e=10

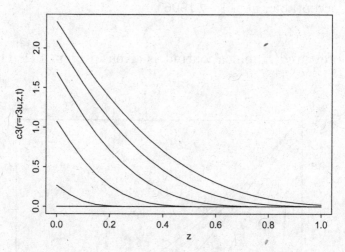

Figure 4.1e: $c_3(r = r_{3u}, z, t)$, n=1, c1e=10

To observe the effect of the nonlinear reaction term in eq. (4.1), the main program of Listing 4.1 is executed with n=2, c1e=10.

```
#
# Select case
  ncase=2;
```

```
if(ncase==1){n=1;}
if(ncase==2){n=2;}
```

Abbreviated output is in Table 4.2.

```
[1] 6
```

```
[1] 274
```

```
ncall = 1865
```

t	z	c1(z,t)	c2l(r,z,t)	c2u(r,z,t)
t	z	c1(z,t)	c3l(r,z,t)	c3u(r,z,t)
0.00	0.00	10.0000	0.0000	0.0000
0.00	0.00	10.0000	0.0000	0.0000
0.00	0.10	0.0000	0.0000	0.0000
0.00	0.10	0.0000	0.0000	0.0000
0.00	0.20	0.0000	0.0000	0.0000
0.00	0.20	0.0000	0.0000	0.0000
0.00	0.30	0.0000	0.0000	0.0000
0.00	0.30	0.0000	0.0000	0.0000
0.00	0.40	0.0000	0.0000	0.0000
0.00	0.40	0.0000	0.0000	0.0000
0.00	0.50	0.0000	0.0000	0.0000
0.00	0.50	0.0000	0.0000	0.0000
0.00	0.60	0.0000	0.0000	0.0000
0.00	0.60	0.0000	0.0000	0.0000
0.00	0.70	0.0000	0.0000	0.0000

```
0.00   0.70   0.0000      0.0000      0.0000

0.00   0.80   0.0000      0.0000      0.0000
0.00   0.80   0.0000      0.0000      0.0000

0.00   0.90   0.0000      0.0000      0.0000
0.00   0.90   0.0000      0.0000      0.0000

0.00   1.00   0.0000      0.0000      0.0000
0.00   1.00   0.0000      0.0000      0.0000
                 .                         .
                 .                         .
                 .                         .
```

Output for t = 0.40,...,1.6 removed

```
                 .                         .
                 .                         .
                 .                         .
```

t	z	c1(z,t)	c21(r,z,t)	c2u(r,z,t)
t	z	c1(z,t)	c31(r,z,t)	c3u(r,z,t)
2.00	0.00	10.0000	9.9902	3.1994
2.00	0.00	10.0000	3.1946	1.4853
2.00	0.10	8.2787	8.2707	2.7583
2.00	0.10	8.2787	2.7544	1.3447
2.00	0.20	6.8195	6.8129	2.3376
2.00	0.20	6.8195	2.3345	1.1851
2.00	0.30	5.5938	5.5883	1.9319
2.00	0.30	5.5938	1.9295	1.0022
2.00	0.40	4.5403	4.5356	1.5374
2.00	0.40	4.5403	1.5355	0.7980

2.00	0.50	3.6181	3.6141	1.1648
2.00	0.50	3.6181	1.1634	0.5878
2.00	0.60	2.8084	2.8050	0.8334
2.00	0.60	2.8084	0.8323	0.3950
2.00	0.70	2.1072	2.1045	0.5598
2.00	0.70	2.1072	0.5590	0.2400
2.00	0.80	1.5164	1.5142	0.3507
2.00	0.80	1.5164	0.3501	0.1307
2.00	0.90	1.0370	1.0354	0.2029
2.00	0.90	1.0370	0.2026	0.0629
2.00	1.00	0.6742	0.6730	0.1101
2.00	1.00	0.6742	0.1099	0.0278

Table 4.2: Solution to eqs. (3.1a,b), (4.1), n=2, c1e=10

Abbreviated graphical output is given in Figs. 4.2a,d,e.

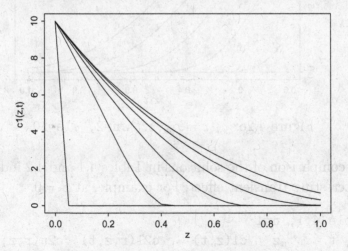

Figure 4.2a: $c_1(z, t)$, n=2, c1e=10

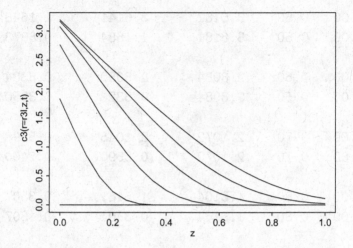

Figure 4.2d: $c_3(r = r_{3l}, z, t)$, n=2, c1e=10

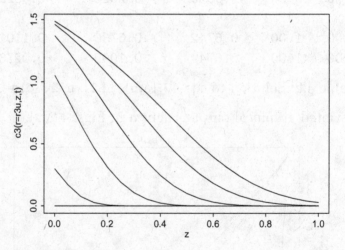

Figure 4.2e: $c_3(r = r_{3u}, z, t)$, n=2, c1e=10

A comparison of the solutions in Tables 4.1 and 4.2 indicates an interesting transient effect. For example, at $z = 0$,

```
Table 4.1, n=1
     t        z    c1(z,t)     c2l(r,z,t)   c2u(r,z,t)
     t        z    c1(z,t)     c3l(r,z,t)   c3u(r,z,t)
  2.00     0.00    10.0000       9.9914       4.0901
```

2.00	0.00	10.0000	4.0860	2.3333

Table 4.2, n=2

t	z	c1(z,t)	c2l(r,z,t)	c2u(r,z,t)
t	z	c1(z,t)	c3l(r,z,t)	c3u(r,z,t)
2.00	0.00	10.0000	9.9902	3.1994
2.00	0.00	10.0000	3.1946	1.4853

The solutions are smaller for n=2 as expected since the second order reaction has a stronger effect than the first order reaction with n=1.

However, this result is reversed at $z = 0.5$.

Table 4.1, n=1

t	z	c1(z,t)	c2l(r,z,t)	c2u(r,z,t)
t	z	c1(z,t)	c3l(r,z,t)	c3u(r,z,t)
2.00	0.50	3.6190	3.6149	1.0926
2.00	0.50	3.6190	1.0911	0.4737

Table 4.2, n=2

t	z	c1(z,t)	c2l(r,z,t)	c2u(r,z,t)
t	z	c1(z,t)	c3l(r,z,t)	c3u(r,z,t)
2.00	0.50	3.6181	3.6141	1.1648
2.00	0.50	3.6181	1.1634	0.5878

This is a transient effect and if the final time is increased from $t_f = 2$ to $t_f = 10$, the concentrations for n=2 at $z = 0.5$ are lower than for n=1.

Table 4.1, n=1, tf=10

t	z	c1(z,t)	c2l(r,z,t)	c2u(r,z,t)
t	z	c1(z,t)	c3l(r,z,t)	c3u(r,z,t)
10.00	0.50	4.5076	4.5040	2.0017
10.00	0.50	4.5076	1.9999	1.2303

Table 4.2, n=2, tf=10

t	z	c1(z,t)	c2l(r,z,t)	c2u(r,z,t)

t	z	c1(z,t)	c31(r,z,t)	c3u(r,z,t)
2.00	0.50	3.6181	3.6141	1.1648
2.00	0.50	3.6181	1.1634	0.5877

This comparison of the solutions for n=1,2 illustrates the straightforward numerical analysis of the effect of the order of the reaction term in eq. (4.1) for the linear and numerical cases. A more detailed understanding of the responses in Figs. 4.1 and 4.2 could be obtained by computing and plotting all of the individual RHS terms and the LHS deriavtives in t of eqs. (3.1a,b), (4.1). This is possible by using the PDE solutions for computing the various terms.

The next example for a nonlinear three PDE model follows.

(4.3) Second Nonlinear PDE Model

The nonlinear model includes diffusivities $D_2 = D_2(c_2(r, z, t))$, $D_3 = D_3(c_3(r, z, t))$ in eqs. (3.1b,c). A restatement of eqs. (3.1) with these nonlinearities follows.

$$\frac{\partial c_1(z,t)}{\partial t} = -v_{c1}\frac{\partial c_1(z,t)}{\partial z} - \frac{2}{r_1}(k_{12f}c_1(z,t) - k_{12r}c_2(r = r_1, z, t))$$

$$(4.2a)$$

Eqs. (4.2b,c) include the nonlinearities[1].

---

[1]For example, the nonlinear diffusion term in eq. (4.1b) is given by

$$\frac{1}{r}\frac{\partial}{\partial r}\left(rD_2(c_2(r,z,t))\frac{\partial c_2(r,z,t)}{\partial r}\right) =$$

$$D_2(c_2(r,z,t))\left(\frac{\partial^2 c_2(r,z,t)}{\partial r^2} + \frac{1}{r}\frac{\partial c_2(r,z,t)}{\partial r}\right)$$

$$+\frac{\partial D_2(c_2(r,z,t))}{\partial c_2(r,z,t)}\left(\frac{\partial c_2(r,z,t)}{\partial r}\right)^2$$

$$\frac{\partial c_2(r,z,t)}{\partial t} = D_2(c_2(r,z,t)) \left(\frac{\partial^2 c_2(r,z,t)}{\partial r^2} + \frac{1}{r}\frac{\partial c_2(r,z,t)}{\partial r} \right)$$

$$+ \frac{\partial D_2(c_2(r,z,t))}{\partial c_2(r,z,t)} \left(\frac{\partial c_2(r,z,t)}{\partial r} \right)^2 \qquad (4.2b)$$

$$\frac{\partial c_3(r,z,t)}{\partial t} = D_3(c_3(r,z,t)) \left(\frac{\partial^2 c_3(r,z,t)}{\partial r^2} + \frac{1}{r}\frac{\partial c_3(r,z,t)}{\partial r} \right)$$

$$+ \frac{\partial D_3(c_3(r,z,t))}{\partial c_3(r,z,t)} \left(\frac{\partial c_3(r,z,t)}{\partial r} \right)^2 - k_1 c_3(r,z,t) \qquad (4.2c)$$

The diffusivities are a function of concentration (and possibly (r, z, t)).

$$D_2 = D_2(c_2(r,z,t), r, z, t) = D_2(a_0 + a_1 c_2(r,z,t)) \qquad (4.3a)$$

$$D_3 = D_3(c_3(r,z,t), r, z, t) = D_3(a_0 + a_1 c_3(r,z,t)) \qquad (4.3b)$$

(the coefficients a_0, a_1 could be different for the membrane and tissue).

The main program for eqs. (4.2) follows.

(4.3.1) Main program

This main program includes the nonlinear diffusion of eqs. (4.2).

```
#
# Three PDE nonlinear BBB model
#
# Delete previous workspaces
  rm(list=ls(all=TRUE))
#
# Access ODE integrator
  library("deSolve");
#
# Access functions for numerical solution
  setwd("f:/BBB/chap4");
```

```
  source("pde1b.R");
  source("dss004.R");
  source("dss044.R");
  source("van1.R");
#
# Model parameters
      u10=0;        u20=0;
      u30=0;        v_c1=1;
      c1e=1;         k1=1;
 D2=1.0e-06; D3=1.0e-06;
     k12f=1;       k12r=1;
     k23f=1;       k23r=1;
#
# Select case
  ncase=1;
#
# Linear
  if(ncase==1){a0=1;a1=0};
#
# Nonlinear
  if(ncase==2){a0=1;a1=10};
#
# Initial condition
  nz=21;nr=6;
  nznr=nz*nr;
  u0=rep(0,nz+2*nznr);
  for(iz in 1:nz){
    u0[iz]=u10;
  for(ir in 1:nr){
    izir=(iz-1)*nr+ir;
    u0[nz+izir]       =u20;
    u0[nz+izir+nznr]=u30;
  }
  }
```

```
#
# Grid in z
  zl=0;zu=1;
  z=seq(from=zl,to=zu,by=(zu-zl)/(nz-1));
#
# Grid in r for c2
  r2l=1.0e-03;r2u=2.0e-03;
  r2=seq(from=r2l,to=r2u,by=(r2u-r2l)/(nr-1));
#
# Grid in r for c3
  r3l=2.0e-03;r3u=3.0e-03;
  r3=seq(from=r3l,to=r3u,by=(r3u-r3l)/(nr-1));
#
# Interval in t
  t0=0;tf=2;nout=6;
  tout=seq(from=t0,to=tf,by=(tf-t0)/(nout-1));
  ncall=0;
#
# ODE integration
  out=lsodes(y=u0,times=tout,func=pde1b,
      sparsetype="sparseint",rtol=1e-6,
      atol=1e-6,maxord=5);
  nrow(out)
  ncol(out)
#
# Store solution
  c1 =matrix(0,nrow=nz,ncol=nout);
  c2l=matrix(0,nrow=nz,ncol=nout);
  c2u=matrix(0,nrow=nz,ncol=nout);
  c3l=matrix(0,nrow=nz,ncol=nout);
  c3u=matrix(0,nrow=nz,ncol=nout);
  t=rep(0,nout);
  for(it in 1:nout){
  for(iz in 1:nz){
```

```
      c1[iz,it]=out[it,iz+1];
    c2l[iz,it]=out[it,nz+1+(iz-1)*nr+1];
    c2u[iz,it]=out[it,nz+1+(iz-1)*nr+nr];
    c3l[iz,it]=out[it,nz+1+(iz-1)*nr+1+nznr];
    c3u[iz,it]=out[it,nz+1+(iz-1)*nr+nr+nznr];
          t[it]=out[it,1];
  }
  c1[1,it]=c1e;
  }
#
# Display ncall
  cat(sprintf("\n\n ncall = %2d",ncall));
#
# Display numerical solution
  for(it in 1:nout){
  cat(sprintf(
    "\n\n       t       z    c1(z,t)      c2l(r,z,t)
    c2u(r,z,t)"));
  cat(sprintf(
      "\n       t      z    c1(z,t)      c3l(r,z,t)
      c3u(r,z,t)"));
  izv=seq(from=1,to=nz,by=2);
  for(iz in izv){
    cat(sprintf(
      "\n%7.2f%7.2f%10.4f%12.4f%12.4f",
      t[it],z[iz],c1[iz,it],c2l[iz,it],
      c2u[iz,it]));
    cat(sprintf(
      "\n%7.2f%7.2f%10.4f%12.4f%12.4f\n",
      t[it],z[iz],c1[iz,it],c3l[iz,it],
      c3u[iz,it]));
  }
  }
#
```

```
# Plot numerical solutions
#
# c1(z,t)
  par(mfrow=c(1,1));
  matplot(
    x=z,y=c1,type="l",xlab="z",ylab="c1(z,t)",
    xlim=c(zl,zu),lty=1,main="",lwd=2,col="black");
#
# c2(r=r2l,z,t)
  par(mfrow=c(1,1));
  matplot(
    x=z,y=c2l,type="l",xlab="z",ylab="c2(r=r2l,z,t)",
    xlim=c(zl,zu),lty=1,main="",lwd=2,col="black");
#
# c2(r=r2u,z,t)
  par(mfrow=c(1,1));
  matplot(
    x=z,y=c2u,type="l",xlab="z",ylab="c2(r=r2u,z,t)",
    xlim=c(zl,zu),lty=1,main="",lwd=2,col="black");
#
# c3(r=rl3,z,t)
  par(mfrow=c(1,1));
  matplot(
    x=z,y=c3l,type="l",xlab="z",ylab="c3(r=r3l,z,t)",
    xlim=c(zl,zu),lty=1,main="",lwd=2,col="black");
#
# c3(r=ru3,z,t)
  par(mfrow=c(1,1));
  matplot(
    x=z,y=c3u,type="l",xlab="z",ylab="c3(r=r3u,z,t)",
    xlim=c(zl,zu),lty=1,main="",lwd=2,col="black");
```

Listing 4.3: Main program for eqs. (4.2)

Listing 4.3 is similar to Listing 3.1 so the differences are emphasized.

- pde1b is the ODE/MOL routine.

```
#
# Access functions for numerical solution
  setwd("f:/BBB/chap4");
  source("pde1b.R");
  source("dss004.R");
  source("dss044.R");
  source("van1.R");
```

- Transfer from the blood to the membrane and tissue is included (via $k_{12f} = 1$ to $k_{23r} = 1$).

```
#
# Model parameters
      u10=0;        u20=0;
      u30=0;        v_c1=1;
      c1e=1;        k1=1;
  D2=1.0e-06; D3=1.0e-06;
      k12f=1;       k12r=1;
      k23f=1;       k23r=1;
```

- Linear or nonlinear diffusion is selected with ncase=1 or ncase=2, respectively.

```
#
# Select case
  ncase=1;
#
# Linear
  if(ncase==1){a0=1;a1=0};
#
# Nonlinear
  if(ncase==2){a0=1;a1=10};
```

a2=10 is selected for `ncase=2` to enhance the nonlinear effect on the solution (discussed subsequently).

- $0 \leq t \leq 2$ is specified as the time interval (discussed with Listing 3.1).

```
#
# Interval in t
  t0=0;tf=2;nout=6;
  tout=seq(from=t0,to=tf,by=(tf-t0)/(nout-1));
  ncall=0;
```

- pde1b is called by lsodes.

```
#
# ODE integration
  out=lsodes(y=u0,times=tout,func=pde1b,
      sparsetype="sparseint",rtol=1e-6,
      atol=1e-6,maxord=5);
  nrow(out)
  ncol(out)
```

pde1b is considered next.

(4.3.2) ODE/MOL routine

pde1b called by lsodes follows.

```
  pde1b=function(t,u,parms){
#
# Function pde1b computes the t derivative
# vector of c1t(z,t), c2t(r,z,t), c3t(r,z,t)
#
# One vector to one vector, two matrices
  c1=rep(0,nz);
  c2=matrix(0,nrow=nz,ncol=nr);
  c3=matrix(0,nrow=nz,ncol=nr);
  for(iz in 1:nz){
```

```
  c1[iz]=u[iz];
for(ir in 1:nr){
  izir=(iz-1)*nr+ir;
  c2[iz,ir]=u[izir+nz];
  c3[iz,ir]=u[izir+nz+nznr];
  }
  }
#
# Boundary condition
  c1[1]=c1e;
#
# c1z
  c1z=vanl(zl,zu,nz,c1,v_c1);
#
# c2r
  c2r=matrix(0,nrow=nz,ncol=nr);
  for(iz in 1:nz){
    c2r[iz,]=dss004(r2l,r2u,nr,c2[iz,]);
  }
#
# c3r
  c3r=matrix(0,nrow=nz,ncol=nr);
  for(iz in 1:nz){
    c3r[iz,]=dss004(r3l,r3u,nr,c3[iz,]);
  }
#
# c2 BCs
  for(iz in 1:nz){
    c2r[iz,1]=
      -(1/D2)*(k12f*c1[iz]    -k12r*c2[iz,1]);
    c2r[iz,nr]=
      -(1/D2)*(k23f*c2[iz,nr]-k23r*c3[iz,1]);
  }
#
```

```
# c3 BCs
  for(iz in 1:nz){
    c3r[iz,1]=
      -(1/D3)*(k23f*c2[iz,nr]-k23r*c3[iz,1]);
    c3r[iz,nr]=0;
  }
#
# c2rr
  nl=2;nu=2;
  c2rr=matrix(0,nrow=nz,ncol=nr);
  for(iz in 1:nz){
    c2rr[iz,]=
      dss044(r2l,r2u,nr,c2[iz,],c2r[iz,],nl,nu);
  }
#
# c3rr
  nl=2;nu=2;
  c3rr=matrix(0,nrow=nz,ncol=nr);
  for(iz in 1:nz){
    c3rr[iz,]=
      dss044(r3l,r3u,nr,c3[iz,],c3r[iz,],nl,nu);
  }
#
# c1t, c2t, c3t
  c1t=rep(0,nz);
  c2t=matrix(0,nrow=nz,ncol=nr);
  c3t=matrix(0,nrow=nz,ncol=nr);
  for(iz in 1:nz){
    c1t[iz]=
      -v_c1*c1z[iz]-(2/r21)*
      (k12f*c1[iz]-k12r*c2[iz,1]);
    for(ir in 1:nr){
      c2t[iz,ir]=
        D2*(a0+a1*c2[iz,ir])*
```

```
          (c2rr[iz,ir]+(1/r2[ir])*c2r[iz,ir])+
          D2*a1*c2r[iz,ir]^2;
      c3t[iz,ir]=
          D3*(a0+a1*c3[iz,ir])*
          (c3rr[iz,ir]+(1/r3[ir])*c3r[iz,ir])+
          D3*a1*c3r[iz,ir]^2-k1*c3[iz,ir];
    }
    }
    c1t[1]=0;
#
# One vector, two matrices to one vector
    ut=rep(0,(nz+nznr));
    for(iz in 1:nz){
      ut[iz]=c1t[iz];
    for(ir in 1:nr){
      izir=(iz-1)*nr+ir;
      ut[nz+izir]      =c2t[iz,ir];
      ut[nz+izir+nznr]=c3t[iz,ir];
    }
    }
#
# Increment calls to pde1b
    ncall<<-ncall+1;
#
# Return derivative vector
    return(list(c(ut)));
}
```

Listing 4.4: ODE/MOL routine for eqs. (4.2)

Listing 4.4 is similar to Listing 3.2 so only the differences are considered.

- The programming up to the calculation of the derivatives $\dfrac{\partial c_1(z,t)}{\partial t}$, $\dfrac{\partial c_2(r,z,t)}{\partial t}$, $\dfrac{\partial c_3(r,z,t)}{\partial t}$ (LHSs of eqs. (4.2)) is the same as in Listing 3.2.

- The programming of eqs. (4.2) steps through r and z in two fors.

```
#
# c1t, c2t, c3t
  c1t=rep(0,nz);
  c2t=matrix(0,nrow=nz,ncol=nr);
  c3t=matrix(0,nrow=nz,ncol=nr);
  for(iz in 1:nz){
    c1t[iz]=
      -v_c1*c1z[iz]-(2/r21)*
      (k12f*c1[iz]-k12r*c2[iz,1]);
  for(ir in 1:nr){
    c2t[iz,ir]=
      D2*(a0+a1*c2[iz,ir])*
      (c2rr[iz,ir]+(1/r2[ir])*c2r[iz,ir])+
      D2*a1*c2r[iz,ir]^2;
    c3t[iz,ir]=
      D3*(a0+a1*c3[iz,ir])*
      (c3rr[iz,ir]+(1/r3[ir])*c3r[iz,ir])+
      D3*a1*c3r[iz,ir]^2-k1*c3[iz,ir];
  }
}
  c1t[1]=0;
```

This programming requires some additional explanation.

– The variable diffusivity $D_2(a_0 + a_1 c_2(r, z, t))$ in eq. (4.3a) is programmed as

```
D2*(a0+a1*c2[iz,ir])
```

– Similarly, the variable diffusivity $D_3(a_0 + a_1 c_3(r, z, t))$ in eq. (4.3b) is programmed as

```
D3*(a0+a1*c3[iz,ir])
```

– The term in eq. (4.2b)

$$\frac{\partial D_2(c_2(r,z,t))}{\partial c_2(r,z,t)} \left(\frac{\partial c_2(r,z,t)}{\partial r} \right)^2$$

$$= D_2 a_1 \left(\frac{\partial c_2(r,z,t)}{\partial r} \right)^2$$

is programmed as

D2*a1*c2r[iz,ir]^2

– Similarly, the term in eq. (4.2c)

$$\frac{\partial D_3(c_3(r,z,t))}{\partial c_3(r,z,t)} \left(\frac{\partial c_3(r,z,t)}{\partial r} \right)^2$$

$$= D_3 a_1 \left(\frac{\partial c_3(r,z,t)}{\partial r} \right)^2$$

is programmed as

D3*a1*c3r[iz,ir]^2

In this way, linear diffusivities are included in eqs. (4.2b,c) with $a_0 = 1, a_1 = 0$, and nonlinear diffusivities are included with $a_0 = 1, a_1 = 10$. In other words, by comparing the solutions for these two cases, the effect of the non-linear term based on a_1 (in eqs. (4.3)) can be ascertained.

These are the only changes in Listing 3.2 to give pde1b in Listing 4.4. Other forms of the nonlinearity can be studied in a similar fashion.

(4.3.3) Model output

The output for ncase=1 ($a_0 = 1, a_1 = 0$) is the same as in Table 3.1 and Figs. 3.1 so it is not repeated here. The output for ncase=2 ($a_0 = 1, a_1 = 10$) follows.

[1] 6

[1] 274

ncall = 1758

t	z	c1(z,t)	c2l(r,z,t)	c2u(r,z,t)
t	z	c1(z,t)	c3l(r,z,t)	c3u(r,z,t)
0.00	0.00	1.0000	0.0000	0.0000
0.00	0.00	1.0000	0.0000	0.0000
0.00	0.10	0.0000	0.0000	0.0000
0.00	0.10	0.0000	0.0000	0.0000
0.00	0.20	0.0000	0.0000	0.0000
0.00	0.20	0.0000	0.0000	0.0000
0.00	0.30	0.0000	0.0000	0.0000
0.00	0.30	0.0000	0.0000	0.0000
0.00	0.40	0.0000	0.0000	0.0000
0.00	0.40	0.0000	0.0000	0.0000
0.00	0.50	0.0000	0.0000	0.0000
0.00	0.50	0.0000	0.0000	0.0000
0.00	0.60	0.0000	0.0000	0.0000
0.00	0.60	0.0000	0.0000	0.0000
0.00	0.70	0.0000	0.0000	0.0000
0.00	0.70	0.0000	0.0000	0.0000
0.00	0.80	0.0000	0.0000	0.0000
0.00	0.80	0.0000	0.0000	0.0000

0.00	0.90	0.0000	0.0000	0.0000
0.00	0.90	0.0000	0.0000	0.0000
0.00	1.00	0.0000	0.0000	0.0000
0.00	1.00	0.0000	0.0000	0.0000

.
.
.

Output for t=0.4,...,1.6 removed

.
.
.

t	z	$c1(z,t)$	$c2l(r,z,t)$	$c2u(r,z,t)$
t	z	$c1(z,t)$	$c3l(r,z,t)$	$c3u(r,z,t)$
2.00	0.00	1.0000	0.9998	0.8609
2.00	0.00	1.0000	0.8608	0.8105
2.00	0.10	0.9625	0.9623	0.8234
2.00	0.10	0.9625	0.8233	0.7730
2.00	0.20	0.9246	0.9245	0.7841
2.00	0.20	0.9246	0.7840	0.7330
2.00	0.30	0.8855	0.8853	0.7405
2.00	0.30	0.8855	0.7404	0.6878
2.00	0.40	0.8434	0.8432	0.6893
2.00	0.40	0.8434	0.6892	0.6334
2.00	0.50	0.7958	0.7956	0.6272
2.00	0.50	0.7958	0.6271	0.5660
2.00	0.60	0.7402	0.7400	0.5524
2.00	0.60	0.7402	0.5522	0.4836

2.00	0.70	0.6748	0.6745	0.4665
2.00	0.70	0.6748	0.4663	0.3885
2.00	0.80	0.5997	0.5994	0.3751
2.00	0.80	0.5997	0.3750	0.2878
2.00	0.90	0.5175	0.5172	0.2863
2.00	0.90	0.5175	0.2861	0.1919
2.00	1.00	0.4324	0.4321	0.2068
2.00	1.00	0.4324	0.2066	0.1109

Table 4.3: Solution to eqs. (4.2), (4.3) with nonlinear diffusion, ncase=2

We can note the following details about this output.

- The dimensions of out from lsodes remain as out(6,21+(12)21+1=273+1=274).

  [1] 6

  [1] 274

- The homogeneous (zero) ICs for eqs. (4.2) programmed in Listing 4.3 are confirmed. Also the BC $c_1(z = 0, t) = c_{1e} = 1$ is confirmed.
- The additional diffusion from $a_1 = 10$ in eqs. (4.2) flattens the profiles in r (the gradients are reduced with the increased diffusion). For example,

  Table 3.1

t	z	c1(z,t)	c2l(r,z,t)	c2u(r,z,t)
t	z	c1(z,t)	c3l(r,z,t)	c3u(r,z,t)

2.00	0.50	0.3663	0.3622	0.1105
2.00	0.50	0.3663	0.1090	0.0474

Table 4.3

t	z	c1(z,t)	c21(r,z,t)	c2u(r,z,t)
t	z	c1(z,t)	c31(r,z,t)	c3u(r,z,t)
2.00	0.50	0.7958	0.7956	0.6272
2.00	0.50	0.7958	0.6271	0.5660

- The computational effort, even with the nonlinear diffusivities, is acceptable.

ncall = 1758

The graphical output is in Figs. 4.3.

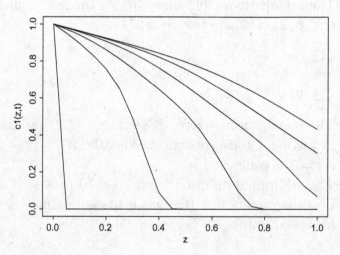

Figure 4.3a: Numerical solution $c_1(z,t)$, $a_0 = 1, a_1 = 10$, ncase=2

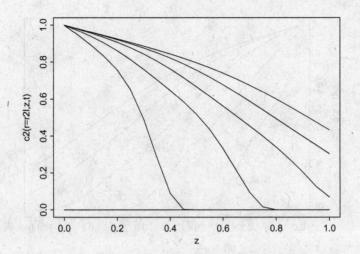

Figure 4.3b: Numerical solution $c_2(r = r_{2l}, z, t)$, $a_0 = 1, a_1 = 10$, ncase=2

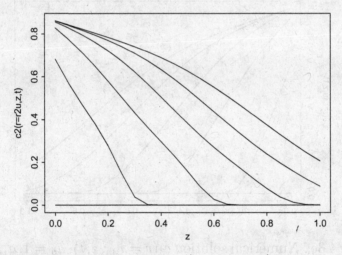

Figure 4.3c: Numerical solution $c_2(r = r_{2u}, z, t)$, $a_0 = 1, a_1 = 10$, ncase=2

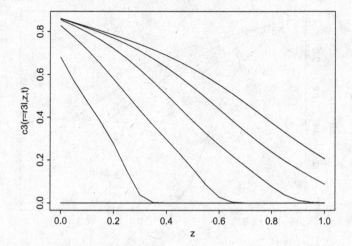

Figure 4.3d: Numerical solution $c_3(r = r_{3l}, z, t)$, $a_0 = 1$, $a_1 = 10$, `ncase=2`

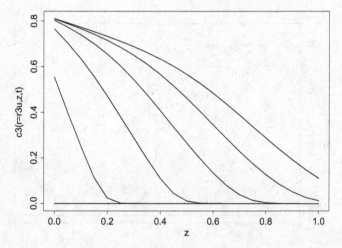

Figure 4.3e: Numerical solution $c_3(r = r_{3u}, z, t)$, $a_0 = 1$, $a_1 = 10$, `ncase=2`

A comparison of Figs. 3.1 and 4.3 indicates that the spatial variation is reduced with the increased diffusivities from the nonlinearity $a_0 = 1$, $a_1 = 10$.

(4.4) Summary and Conclusions

The preceding examples demonstrate that including nonlinearities in the numerical solution is straightforward, which is in contrast to an analytical approach for which this would be relatively difficult. Thus, the model formulation is significantly extended since nonlinearties can be readily included numerically.

In the next chapter, two components are included in the BBB model, which could represent, for example, the simultaneous transfer of O_2 and a nutrient.

Chapter 5

Multicomponent Blood Brain Barrier Models

(5.1) Introduction

The models for the blood brain barrier (BBB) in the preceding chapters pertain to one transferred component with concentrations $c_1(z,t)$ (blood), $c_2(r,z,t)$ (membrane) and $c_3(r,z,t)$ (tissue). In this chapter consideration is given to BBB models with more than one transferred component.

(5.2) Two Component BBB Model

If two components are considered, their concentrations are designated as follows.

Component	Section	Notation
1	blood	$c_{1,1}(z,t)$
1	membrane	$c_{2,1}(r,z,t)$
1	tissue	$c_{3,1}(r,z,t)$
2	blood	$c_{1,2}(z,t)$
2	membrane	$c_{2,2}(r,z,t)$
2	tissue	$c_{3,2}(r,z,t)$

Table 5.1: Notation for two component system

Briefly, the first subscript of each concentration designates the section (blood, membrane, tissue) and the second subscript designates the component.

A mass balance on each section for each component gives the following system of PDEs (see Chapter 1 for detailed PDE derivations).

$$\frac{\partial c_{1,1}(z,t)}{\partial t} = -v_{c1,1}\frac{\partial c_{1,1}(z,t)}{\partial z}$$

$$-\frac{2}{r_1}(k_{12f}c_{1,1}(z,t) - k_{12r}c_{2,1}(r=r_1,z,t)) \tag{5.1a}$$

$$\frac{\partial c_{1,2}(z,t)}{\partial t} = -v_{c1,2}\frac{\partial c_{1,2}(z,t)}{\partial z}$$

$$-\frac{2}{r_1}(k_{12f}c_{1,2}(z,t) - k_{12r}c_{2,2}(r=r_1,z,t)) \tag{5.1b}$$

$$\frac{\partial c_{2,1}(r,z,t)}{\partial t} = D_{2,1}\left(\frac{\partial^2 c_{2,1}(r,z,t)}{\partial r^2} + \frac{1}{r}\frac{\partial c_{2,1}(r,z,t)}{\partial r}\right) \tag{5.2a}$$

$$\frac{\partial c_{2,2}(r,z,t)}{\partial t} = D_{2,2}\left(\frac{\partial^2 c_{2,2}(r,z,t)}{\partial r^2} + \frac{1}{r}\frac{\partial c_{2,2}(r,z,t)}{\partial r}\right) \tag{5.2b}$$

$$\frac{\partial c_{3,1}(r,z,t)}{\partial t} = D_{3,1}\left(\frac{\partial^2 c_{3,1}(r,z,t)}{\partial r^2} + \frac{1}{r}\frac{\partial c_{3,1}(r,z,t)}{\partial r}\right)$$

$$-k_1 c_{3,1}(r,z,t) - k_{12}c_{3,1}(r,z,t)c_{3,2}(r,z,t) \tag{5.3a}$$

$$\frac{\partial c_{3,2}(r,z,t)}{\partial t} = D_{3,2}\left(\frac{\partial^2 c_{3,2}(r,z,t)}{\partial r^2} + \frac{1}{r}\frac{\partial c_{3,2}(r,z,t)}{\partial r}\right)$$

$$-k_1 c_{3,2}(r,z,t) - k_{12}c_{3,1}(r,z,t)c_{3,2}(r,z,t) \tag{5.3b}$$

Eqs. (5.1) to (5.3) follow from eqs. (3.1). The bimolecular reaction rate

$$-k_{12}c_{3,1}(r,z,t)c_{3,2}(r,z,t)$$

is included in eqs. (5.3) that provides a link between the c_1 and c_2 balances, eqs. (5.3).

Eqs. (5.1) are first order in z and t so that each requires one IC and one BC. Eqs. (5.2) and (5.3) are second order in r and first order in t so that each requires one IC and two BCs. These auxiliary conditions are listed next.

IC, BC for eqs. (5.1).

$$c_{1,1}(z,t=0) = c_{10,1}(z); \quad c_{1,1}(z=0,t) = c_{1e,1}(t) \qquad \text{(5.4a,b)}$$

$$c_{1,2}(z,t=0) = c_{10,2}(z); \quad c_{1,2}(z=0,t) = c_{1e,2}(t) \qquad \text{(5.4c,d)}$$

IC, BCs for eqs. (5.2).

$$c_{2,1}(r,z,t=0) = c_{20,1}(r,z) \qquad \text{(5.5a)}$$

$$D_{2,1}\frac{c_{2,1}(r=r_{2l},z,t)}{\partial r} =$$

$$-(k_{12f,1}c_{1,1}(z,t) - k_{12r,1}c_{2,1}(r=r_{2l},z,t)) \qquad \text{(5.5b)}$$

$$D_{2,1}\frac{c_{2,1}(r=r_{2u},z,t)}{\partial r} =$$

$$-(k_{23f,1}c_{2,1}(r=r_{2u},z,t) - k_{23r,1}c_{3,1}(r=r_{2u},z,t)) \qquad \text{(5.5c)}$$

$$c_{2,2}(r,z,t=0) = c_{20,2}(r,z) \qquad \text{(5.5d)}$$

$$D_{2,2}\frac{c_{2,2}(r=r_{2l},z,t)}{\partial r} =$$

$$-(k_{12f,2}c_{1,2}(z,t) - k_{12r,2}c_{2,2}(r=r_{2l},z,t)) \qquad \text{(5.5e)}$$

$$D_{2,2}\frac{c_{2,2}(r = r_{2u}, z, t)}{\partial r} =$$

$$-(k_{23f,2}c_{2,2}(r = r_{2u}, z, t) - k_{23r,2}c_{3,2}(r = r_{2u}, z, t)) \qquad (5.5f)$$

IC, BCs for eqs. (5.3).

$$c_{3,1}(r, z, t = 0) = c_{30,1}(r, z) \qquad (5.6a)$$

$$D_{3,1}\frac{c_{3,1}(r = r_{3l}, z, t)}{\partial r} =$$

$$-(k_{23f,1}c_{2,1}(r = r_{2u}, z, t) - k_{23r,1}c_{3,1}(r = r_{3l}, z, t)) \qquad (5.6b)$$

$$D_{3,1}\frac{c_{3,1}(r = r_{3u}, z, t)}{\partial r} = 0 \qquad (5.6c)$$

$$c_{3,2}(r, z, t = 0) = c_{30,2}(r, z) \qquad (5.6d)$$

$$D_{3,2}\frac{c_{3,2}(r = r_{3l}, z, t)}{\partial r} =$$

$$-(k_{23f,2}c_{2,2}(r = r_{2u}, z, t) - k_{23r,2}c_{3,2}(r = r_{3l}, z, t)) \qquad (5.6e)$$

$$D_{3,2}\frac{c_{3,2}(r = r_{3u}, z, t)}{\partial r} = 0 \qquad (5.6f)$$

A main pogram and a ODE/MOL routine for eqs. (5.1) to (5.6) follow.

(5.2.1) Main program

```
#
# Two component, three PDE BBB model
#
# Delete previous workspaces
  rm(list=ls(all=TRUE))
```

```
#
# Access ODE integrator
  library("deSolve");
#
# Access functions for numerical solution
  setwd("f:/BBB/chap5");
  source("pde1a.R");
  source("dss004.R");
  source("dss044.R");
  source("vanl.R");
#
# Model parameters
#
# First component
        u10_1=0;      u20_1=0;
        u30_1=0;      vc1_1=1;
        c1e_1=1;       k1_1=1;
    D2_1=1.0e-06; D3_1=1.0e-06;
      k12f_1=1;     k12r_1=1;
      k23f_1=1;     k23r_1=1;
#
# Second component
        u10_2=0;      u20_2=0;
        u30_2=0;      vc1_2=1;
        c1e_2=1;       k1_2=1;
    D2_2=1.0e-06; D3_2=1.0e-06;
      k12f_2=1;     k12r_2=1;
      k23f_2=1;     k23r_2=1;
#
# First+second component reaction
      kr_12=1;
#
# Initial condition
  nz=21;nr=6;
```

```
  nznr=nz*nr;nznr2=nz+2*nznr;
  u0=rep(0,2*nznr2);
#
# First component
  nznr=nz*nr;
  for(iz in 1:nz){
    u0[iz]=u10_1;
    for(ir in 1:nr){
    izir=(iz-1)*nr+ir;
    u0[nz+izir]       =u20_1;
    u0[nz+izir+nznr]=u30_1;
  }
  }
#
# Second component
  for(iz in 1:nz){
    u0[iz+nznr2]=u10_2;
    for(ir in 1:nr){
    izir=(iz-1)*nr+ir;
    u0[nz+izir+nznr2]       =u20_2;
    u0[nz+izir+nznr+nznr2]=u30_2;
  }
  }
#
# Grid in z
  zl=0;zu=1;
  z=seq(from=zl,to=zu,by=(zu-zl)/(nz-1));
#
# Grid in r for c2
  r2l=1.0e-03;r2u=2.0e-03;
  r2=seq(from=r2l,to=r2u,by=(r2u-r2l)/(nr-1));
#
# Grid in r for c3
  r3l=2.0e-03;r3u=3.0e-03;
```

```
  r3=seq(from=r31,to=r3u,by=(r3u-r31)/(nr-1));
#
# Interval in t
  t0=0;tf=2;nout=6;
  tout=seq(from=t0,to=tf,by=(tf-t0)/(nout-1));
  ncall=0;
#
# ODE integration
  out=lsodes(y=u0,times=tout,func=pde1a,
      sparsetype="sparseint",rtol=1e-4,
      atol=1e-4,maxord=2);
  nrow(out)
  ncol(out)
#
# Store solutions
#
# First component
  c1_1 =matrix(0,nrow=nz,ncol=nout);
  c2l_1=matrix(0,nrow=nz,ncol=nout);
  c2u_1=matrix(0,nrow=nz,ncol=nout);
  c3l_1=matrix(0,nrow=nz,ncol=nout);
  c3u_1=matrix(0,nrow=nz,ncol=nout);
  t=rep(0,nout);
  for(it in 1:nout){
  for(iz in 1:nz){
    c1_1[iz,it]=
      out[it,iz+1];
    c2l_1[iz,it]=
      out[it,nz+1+(iz-1)*nr+1];
    c2u_1[iz,it]=
      out[it,nz+1+(iz-1)*nr+nr];
    c3l_1[iz,it]=
      out[it,nz+1+(iz-1)*nr+1+nznr];
    c3u_1[iz,it]=
```

```
        out[it,nz+1+(iz-1)*nr+nr+nznr];
      t[it]=out[it,1];
    }
    c1_1[1,it]=c1e_1;
    }
#
# Second component
  c1_2 =matrix(0,nrow=nz,ncol=nout);
  c2l_2=matrix(0,nrow=nz,ncol=nout);
  c2u_2=matrix(0,nrow=nz,ncol=nout);
  c3l_2=matrix(0,nrow=nz,ncol=nout);
  c3u_2=matrix(0,nrow=nz,ncol=nout);
  for(it in 1:nout){
  for(iz in 1:nz){
    c1_2[iz,it]=
      out[it,iz+1+nznr2];
    c2l_2[iz,it]=
      out[it,nz+1+(iz-1)*nr+1+nznr2];
    c2u_2[iz,it]=
      out[it,nz+1+(iz-1)*nr+nr+nznr2];
    c3l_2[iz,it]=
      out[it,nz+1+(iz-1)*nr+1+nznr+nznr2];
    c3u_2[iz,it]=
      out[it,nz+1+(iz-1)*nr+nr+nznr+nznr2];
    }
  c1_2[1,it]=c1e_2;
    }
#
# Display ncall
  cat(sprintf("\n\n ncall = %2d",ncall));
#
# Display numerical solutions
#
# First component
```

```
for(it in 1:nout){
cat(sprintf(
  "\n\n      t        z      c1_1(z,t)
    c2l_1(r,z,t)  c2u_1(r,z,t)"));
cat(sprintf(
    "\n       t        z      c1_1(z,t)
    c3l_1(r,z,t)  c3u_1(r,z,t)"));
izv=seq(from=1,to=nz,by=2);
for(iz in izv){
  cat(sprintf(
    "\n%7.2f%7.2f%12.4f%14.4f%14.4f",
    t[it],z[iz],c1_1[iz,it],
    c2l_1[iz,it],c2u_1[iz,it]));
  cat(sprintf(
    "\n%7.2f%7.2f%12.4f%14.4f%14.4f\n",
    t[it],z[iz],c1_1[iz,it],
    c3l_1[iz,it],c3u_1[iz,it]));
  }
  }
#
# Second component
  for(it in 1:nout){
  cat(sprintf(
    "\n\n      t        z      c1_2(z,t)
    c2l_2(r,z,t)  c2u_2(r,z,t)"));
  cat(sprintf(
      "\n      t        z      c1_2(z,t)
    c3l_2(r,z,t)  c3u_2(r,z,t)"));
  izv=seq(from=1,to=nz,by=2);
  for(iz in izv){
    cat(sprintf(
      "\n%7.2f%7.2f%12.4f%14.4f%14.4f",
      t[it],z[iz],c1_2[iz,it],
      c2l_2[iz,it],c2u_2[iz,it]));
```

```
      cat(sprintf(
        "\n%7.2f%7.2f%12.4f%14.4f%14.4f\n",
        t[it],z[iz],c1_2[iz,it],
        c3l_2[iz,it],c3u_2[iz,it]));
    }
  }
#
# Plot numerical solutions
#
# First component
# c1_1(z,t)
  par(mfrow=c(1,1));
  matplot(
    x=z,y=c1_1,type="l",xlab="z",
    ylab="c1_1(z,t)",xlim=c(zl,zu),
    lty=1,main="",lwd=2,col="black");
#
# c2_1(r=r2l,z,t)
  par(mfrow=c(1,1));
  matplot(
    x=z,y=c2l_1,type="l",xlab="z",
    ylab="c2_1(r=r2l,z,t)",xlim=c(zl,zu),
    lty=1,main="",lwd=2,col="black");
#
# c2_1(r=r2u,z,t)2(r=r2u,z,t)
  par(mfrow=c(1,1));
  matplot(
    x=z,y=c2u_1,type="l",xlab="z",
    ylab="c2_1(r=r2u,z,t)",xlim=c(zl,zu),
    lty=1,main="",lwd=2,col="black");
#
# c3_1(r=rl3,z,t)
  par(mfrow=c(1,1));
  matplot(
```

```
    x=z,y=c3l_1,type="l",xlab="z",
    ylab="c3_1(r=r3l,z,t)",xlim=c(zl,zu),
    lty=1,main="",lwd=2,col="black");
#
# c3_1(r=ru3,z,t)
  par(mfrow=c(1,1));
  matplot(
    x=z,y=c3u_1,type="l",xlab="z",
    ylab="c3_1(r=r3u,z,t)",xlim=c(zl,zu),
    lty=1,main="",lwd=2,col="black");
#
# Second component
# c1_2(z,t)
  par(mfrow=c(1,1));
  matplot(
    x=z,y=c1_2,type="l",xlab="z",
    ylab="c1_2(z,t)",xlim=c(zl,zu),
    lty=1,main="",lwd=2,col="black");
#
# c2_2(r=r2l,z,t)
  par(mfrow=c(1,1));
  matplot(
    x=z,y=c2l_2,type="l",xlab="z",
    ylab="c2_2(r=r2l,z,t)",xlim=c(zl,zu),
    lty=1,main="",lwd=2,col="black");
#
# c2_2(r=r2u,z,t)2(r=r2u,z,t)
  par(mfrow=c(1,1));
  matplot(
    x=z,y=c2u_2,type="l",xlab="z",
    ylab="c2_2(r=r2u,z,t)",xlim=c(zl,zu),
    lty=1,main="",lwd=2,col="black");
#
# c3_2(r=rl3,z,t)
```

```
  par(mfrow=c(1,1));
  matplot(
    x=z,y=c3l_2,type="l",xlab="z",
    ylab="c3_2(r=r3l,z,t)",xlim=c(zl,zu),
    lty=1,main="",lwd=2,col="black");
#
# c3_2(r=ru3,z,t)
  par(mfrow=c(1,1));
  matplot(
    x=z,y=c3u_2,type="l",xlab="z",
    ylab="c3_2(r=r3u,z,t)",xlim=c(zl,zu),
    lty=1,main="",lwd=2,col="black");
#
# Compute and display c1*c2
#
# r=r3l
# Compute kr_12*c3_1*c3_2
  c31c32=matrix(0,nrow=nz,ncol=nout);
  for(it in 1:nout){
    c31c32[,it]=kr_12*c3l_1[,it]*c3l_2[,it];
  }
#
# Plot kr_12*c3_1*c3_2
  par(mfrow=c(1,1));
  matplot(
    x=z,c31c32,type="l",xlab="z",
    ylab="k_12*c1*c2,r=r3l",xlim=c(zl,zu),
    lty=1,main="",lwd=2,col="black");
#
# r=r3u
# Compute kr_12*c3_1*c3_2
  for(it in 1:nout){
    c31c32[,it]=kr_12*c3u_1[,it]*c3u_2[,it];
  }
```

```
#
# Plot kr_12*c3_1*c3_2
  par(mfrow=c(1,1));
  matplot(
    x=z,c31c32,type="l",xlab="z",
    ylab="k_12*c1*c2,r=r3u",xlim=c(zl,zu),
    lty=1,main="",lwd=2,col="black");
```

Listing 5.1: Main program for eqs. (5.1) to (5.6)

We can note the following details about Listing 5.1.

- Previous workspaces are deleted.

```
  #
  # Two component, three PDE BBB model
  #
  # Delete previous workspaces
    rm(list=ls(all=TRUE))
```

- The R ODE integrator library deSolve is accessed. Then the directory with the files for the solution of eqs. (5.1) to (5.6) is designated. Note that setwd (set working directory) uses / rather than the usual \.

```
  #
  # Access ODE integrator
    library("deSolve");
  #
  # Access functions for numerical solution
    setwd("f:/BBB/chap5");
    source("pde1a.R");
    source("dss004.R");
    source("dss044.R");
    source("van1.R");
```

pde1a.R is the routine for the method of lines (MOL) approximation of PDEs (5.1) (discussed subsequently). dss004, dss044 (Differentiation in Space Subroutine) are library routines for calculating first and second derivatives in r by finite differences (FDs). vanl is a library routine for calculating first derivatives in z. The coding and use of these routines is discussed subsequently.

- The parameters of eqs. (5.1) to (5.6) are defined numerically for the two components.

```
#
# Model parameters
#
# First component
        u10_1=0;        u20_1=0;
        u30_1=0;        vc1_1=1;
        c1e_1=1;        k1_1=1;
    D2_1=1.0e-06; D3_1=1.0e-06;
        k12f_1=1;       k12r_1=1;
        k23f_1=1;       k23r_1=1;
#
# Second component
        u10_2=0;        u20_2=0;
        u30_2=0;        vc1_2=1;
        c1e_2=1;        k1_2=1;
    D2_2=1.0e-06; D3_2=1.0e-06;
        k12f_2=1;       k12r_2=1;
        k23f_2=1;       k23r_2=1;
#
# First+second component reaction
      kr_12=1;
```

u10_1 to k23r_1 are for component one and u10_2 to k23r_2 are for component two. kr_12 is a biomeolecular reaction rate constant that links the equations fothe the two components.

Initially, the parameters are the same numerically for the two components for testing of the programming (the solutions for the two components are the same and this can be used as a check of the programming). These parameters are used in the ODE/MOL routine **pde1a** which has the programming of eqs. (5.1) to (5.6).

- The IC vector is placed in u0.

```
#
# Initial condition
  nz=21;nr=6;
  nznr=nz*nr;nznr2=nz+2*nznr;
  u0=rep(0,2*nznr2);
#
# First component
  nznr=nz*nr;
  for(iz in 1:nz){
    u0[iz]=u10_1;
    for(ir in 1:nr){
    izir=(iz-1)*nr+ir;
    u0[nz+izir]      =u20_1;
    u0[nz+izir+nznr]=u30_1;
    }
  }
#
# Second component
  for(iz in 1:nz){
    u0[iz+nznr2]=u10_2;
    for(ir in 1:nr){
    izir=(iz-1)*nr+ir;
    u0[nz+izir+nznr2]      =u20_2;
    u0[nz+izir+nznr+nznr2]=u30_2;
```

```
}
}
```

u0 for the dimensions nz=21, nr=6 has a total number of MOL/ODEs (length) 2*(nz+2*nz*nr) = 2*(nz+2*nznr) = 2*(21+2*21*6)=546. That is, the model consists of two components, each with three PDEs, two PDEs with nz=21 points in z (eqs. (5.1a,b)) and four PDEs with nz*nr=21*6 points in z and r (eqs. (5.2a,b), (5.3a,b)).

The three IC values u10_1, u20_1, u30_1 are for eqs. (5.1a), (5.2a), (5.3a), respectively, that is ICs (5.4a), (5.5a), (5.6a)

The three IC values u10_2, u20_2, u30_2 are for eqs. (5.1b), (5.2b), (5.3b), respectively, that is ICs (5.4c), (5.5d), (5.6d)

- The grid in z, the same for components one, two, is defined for $z_l = 0 \leq z \leq z_u = 1$, so that $z = 0, 0.05, \ldots, 1$.

```
#
# Grid in z
  zl=0;zu=1;
  z=seq(from=zl,to=zu,by=(zu-zl)/(nz-1));
```

- The grid in r for the membrane, r_2, for components one, two, is defined for $r_{2l} = 0.001 \leq r_2 \leq r_{2u} = 0.002$, so that $r_2 = 0.001, 0.0012, \ldots, 0.002$.

```
#
# Grid in r for c2
  r2l=1.0e-03;r2u=2.0e-03;
  r2=seq(from=r2l,to=r2u,by=(r2u-r2l)/(nr-1));
```

- The grid in r for the tissue, r_3, for components one, two, is defined for $r_{3l} = 0.002 \leq r_3 \leq r_{3u} = 0.003$, so that $r_3 = 0.002, 0.0022, \ldots, 0.003$.

```
#
# Grid in r for c3
  r3l=2.0e-03;r3u=3.0e-03;
  r3=seq(from=r3l,to=r3u,by=(r3u-r3l)/(nr-1));
```

- An interval in t of 6 points is defined for $0 \leq t \leq 2$ so that tout=0,0.4,...,2.

```
#
# Interval in t
  t0=0;tf=2;nout=6;
  tout=seq(from=t0,to=tf,by=(tf-t0)/(nout-1));
  ncall=0;
```

The counter for the calls to the ODE/MOL routine pde1a is also initialized.

- The system of 546 MOL/ODEs is integrated by the library integrator lsodes (available in deSolve). As expected, the inputs to lsodes are the ODE function, pde1a, the IC vector u0, and the vector of output values of t, tout. The length of u0 (e.g., 546) informs lsodes how many ODEs are to be integrated. func,y,times are reserved names.

```
#
# ODE integration
  out=lsodes(y=u0,times=tout,func=pde1a,
      sparsetype="sparseint",rtol=1e-4,
      atol=1e-4,maxord=2);
  nrow(out)
  ncol(out)
```

The numerical solution to the ODEs is returned in matrix out. In this case, out has the dimensions $nout \times 2(nz + 2(nz)(nr)) + 1 = 6 \times 2(21 + (2)(21)(6)) + 1 = 547$, which are confirmed by the output from `nrow(out)`,`ncol(out)` (included in the numerical output considered subsequently).

The offset $546+1 = 547$ is required since the first element of each column has the output t (also in `tout`), and the $2, \ldots, 546+1 = 2, \ldots, 547$ column elements have the 546 ODE solutions.

The size of this initial value ODE system with a Jacobian matrix of $546^2 = 298116$ elements indicates the advantage of using the sparse matrix facility of `lsodes`. To reduce the computational time, the error tolerances were increased from `rtol=1e-6`, `atol=1e-6` (the default values) to `rtol=1e-4`, `atol=1e-4`. Also, the maximum order of the ODE integrator was reduced from `maxord=5` (the default value) to `maxord=2`. This change in the maximum order is based partly on the observation that for `maxord=2`, the stability region of the BDF integrator is the entire left complex plane [1] and for `maxord > 2`, a portion of the left complex plane is an unstable region so that the compute times can be much higher and/or the integration can fail.

- Selected component one solutions returned by `out` are placed in matrices `c1_1`, `c2l_1`, `c2u_1`, `c3l_1`, `c3u_1`. t is also placed in vector `t`.

```
#
# Store solutions
#
# First component
  c1_1 =matrix(0,nrow=nz,ncol=nout);
  c2l_1=matrix(0,nrow=nz,ncol=nout);
  c2u_1=matrix(0,nrow=nz,ncol=nout);
  c3l_1=matrix(0,nrow=nz,ncol=nout);
  c3u_1=matrix(0,nrow=nz,ncol=nout);
  t=rep(0,nout);
```

```
for(it in 1:nout){
for(iz in 1:nz){
  c1_1[iz,it]=
    out[it,iz+1];
  c21_1[iz,it]=
    out[it,nz+1+(iz-1)*nr+1];
  c2u_1[iz,it]=
    out[it,nz+1+(iz-1)*nr+nr];
  c31_1[iz,it]=
    out[it,nz+1+(iz-1)*nr+1+nznr];
  c3u_1[iz,it]=
    out[it,nz+1+(iz-1)*nr+nr+nznr];
  t[it]=out[it,1];
}
c1_1[1,it]=c1e_1;
}
```

nznr=nz*nr was set previously. c1_1[1,it]=c1e_1 sets
BC (5.4b) since this value is not returned by lsodes (it is
defined algebraically in pde1a and not by the integration
of an ODE at $z = 0$).

- Similarly, selected component two solutions returned by
  out are placed in matrices c1_2, c21_2, c2u_2, c31_2,
  c3u_2.

```
#
# Second component
  c1_2 =matrix(0,nrow=nz,ncol=nout);
  c21_2=matrix(0,nrow=nz,ncol=nout);
  c2u_2=matrix(0,nrow=nz,ncol=nout);
  c31_2=matrix(0,nrow=nz,ncol=nout);
  c3u_2=matrix(0,nrow=nz,ncol=nout);
  for(it in 1:nout){
  for(iz in 1:nz){
    c1_2[iz,it]=
```

```
        out[it,iz+1+nznr2];
      c2l_2[iz,it]=
        out[it,nz+1+(iz-1)*nr+1+nznr2];
      c2u_2[iz,it]=
        out[it,nz+1+(iz-1)*nr+nr+nznr2];
      c3l_2[iz,it]=
        out[it,nz+1+(iz-1)*nr+1+nznr+nznr2];
      c3u_2[iz,it]=
        out[it,nz+1+(iz-1)*nr+nr+nznr+nznr2];
    }
    c1_2[1,it]=c1e_2;
    }
```

nznr2=nz+2*nz*nr was set previously. c1_2[1,it]=
c1e_2 sets BC (5.4d). Note that the solutions for component two have indices for the corresponding component one solutions offset by nz+2*nz*nr=nznr2=273, e.g.,
c1_2[iz,it]=out[it,iz+1+nznr2].

- The number of calls to pde1a is displayed at the end of
the solution.

```
#
# Display ncall
  cat(sprintf("\n\n ncall = %2d",ncall));
```

- Selected component one numerical solutions are displayed.

```
#
# First component
  for(it in 1:nout){
  cat(sprintf(
    "\n\n      t        z      c1_1(z,t)
    c2l_1(r,z,t)   c2u_1(r,z,t)"));
  cat(sprintf(
    "\n        t        z      c1_1(z,t)
```

```
        c31_1(r,z,t) · c3u_1(r,z,t)")));
  izv=seq(from=1,to=nz,by=2);
  for(iz in izv){
    cat(sprintf(
      "\n%7.2f%7.2f%12.4f%14.4f%14.4f",
      t[it],z[iz],c1_1[iz,it],
      c21_1[iz,it],c2u_1[iz,it]));
    cat(sprintf(
      "\n%7.2f%7.2f%12.4f%14.4f%14.4f\n",
      t[it],z[iz],c1_1[iz,it],
      c31_1[iz,it],c3u_1[iz,it]));
  }
}
```

Every second value in z appears with by=2.

- Similarly, selected component two numerical solutions are displayed.

```
#
# Second component
  for(it in 1:nout){
  cat(sprintf(
    "\n\n      t       z      c1_2(z,t)
      c21_2(r,z,t)   c2u_2(r,z,t)")));
  cat(sprintf(
    "\n       t       z      c1_2(z,t)
      c31_2(r,z,t)   c3u_2(r,z,t)")));
  izv=seq(from=1,to=nz,by=2);
  for(iz in izv){
    cat(sprintf(
      "\n%7.2f%7.2f%12.4f%14.4f%14.4f",
      t[it],z[iz],c1_2[iz,it],
      c21_2[iz,it],c2u_2[iz,it]));
    cat(sprintf(
```

```
      "\n%7.2f%7.2f%12.4f%14.4f%14.4f\n",
      t[it],z[iz],c1_2[iz,it],
      c3l_2[iz,it],c3u_2[iz,it]));
  }
  }
```

- The component one solutions are plotted (total of five graphs).

```
#
# Plot numerical solutions
#
# First component
# c1_1(z,t)
  par(mfrow=c(1,1));
  matplot(
    x=z,y=c1_1,type="l",xlab="z",
    ylab="c1_1(z,t)",xlim=c(zl,zu),
    lty=1,main="",lwd=2,col="black");



#
# c3_1(r=ru3,z,t)
  par(mfrow=c(1,1));
  matplot(
    x=z,y=c3u_1,type="l",xlab="z",
    ylab="c3_1(r=r3u,z,t)",xlim=c(zl,zu),
    lty=1,main="",lwd=2,col="black");
```

- The component two solutions are plotted (total of five graphs).

```
#
# Second component
# c1_2(z,t)
```

```
   par(mfrow=c(1,1));
   matplot(
     x=z,y=c1_2,type="l",xlab="z",
     ylab="c1_2(z,t)",xlim=c(zl,zu),
     lty=1,main="",lwd=2,col="black");
```

.
.
.

```
#
# c3_2(r=ru3,z,t)
   par(mfrow=c(1,1));
   matplot(
     x=z,y=c3u_2,type="l",xlab="z",
     ylab="c3_2(r=r3u,z,t)",xlim=c(zl,zu),
     lty=1,main="",lwd=2,col="black");
```

- The product function kr_12*c31_1[,it]*c31_2[,it] is computed and plotted.

```
#
# Compute and display c1*c2
#
# r=r3l
# Compute kr_12*c3_1*c3_2
   c31c32=matrix(0,nrow=nz,ncol=nout);
   for(it in 1:nout){
     c31c32[,it]=kr_12*c31_1[,it]*c31_2[,it];
   }
#
# Plot kr_12*c3_1*c3_2
   par(mfrow=c(1,1));
   matplot(
     x=z,c31c32,type="l",xlab="z",
     ylab="k_12*c1*c2,r=r3l",xlim=c(zl,zu),
     lty=1,main="",lwd=2,col="black");
```

- Similarly, the product function `kr_23*c3u_1[,it]*` `c3u_2[,it]` is computed and plotted.

```
#
# r=r3u
# Compute kr_12*c3_1*c3_2
  for(it in 1:nout){
    c31c32[,it]=kr_12*c3u_1[,it]*c3u_2[,it];
  }
#
# Plot kr_12*c3_1*c3_2
  par(mfrow=c(1,1));
  matplot(
    x=z,c31c32,type="l",xlab="z",
    ylab="k_12*c1*c2,r=r3u",xlim=c(zl,zu),
    lty=1,main="",lwd=2,col="black");
```

This completes the discussion of the main program in Listing 5.1. The ODE/MOL routine, `pde1a`, called by `lsodes` follows.

(5.2.2) ODE/MOL routine

```
  pde1a=function(t,u,parms){
#
# Function pde1a computes the t derivative
# vector of two components
#
# c1t_1(z,t), c2t_1(r,z,t), c3t_1(r,z,t)
#
# c1t_2(z,t), c2t_2(r,z,t), c3t_2(r,z,t)
#
# First component, one vector to one vector,
# two matrices
  c1_1=rep(0,nz);
  c2_1=matrix(0,nrow=nz,ncol=nr);
```

```
  c3_1=matrix(0,nrow=nz,ncol=nr);
  for(iz in 1:nz){
    c1_1[iz]=u[iz];
  for(ir in 1:nr){
    izir=(iz-1)*nr+ir;
    c2_1[iz,ir]=u[izir+nz];
    c3_1[iz,ir]=u[izir+nz+nznr];
  }
  }
#
# Second component, one vector to one vector,
# two matrices
  c1_2=rep(0,nz);
  c2_2=matrix(0,nrow=nz,ncol=nr);
  c3_2=matrix(0,nrow=nz,ncol=nr);
  for(iz in 1:nz){
    c1_2[iz]=u[iz+nznr2];
  for(ir in 1:nr){
    izir=(iz-1)*nr+ir;
    c2_2[iz,ir]=u[izir+nz+nznr2];
    c3_2[iz,ir]=u[izir+nz+nznr+nznr2];
  }
  }
#
# First component ODE/MOL
#
# Boundary condition
  c1_1[1]=c1e_1;
#
# c1z
  c1z_1=vanl(zl,zu,nz,c1_1,vc1_1);
#
# c2r
  c2r_1=matrix(0,nrow=nz,ncol=nr);
```

```
  for(iz in 1:nz){
    c2r_1[iz,]=dss004(r2l,r2u,nr,c2_1[iz,]);
  }
#
# c3r
  c3r_1=matrix(0,nrow=nz,ncol=nr);
  for(iz in 1:nz){
    c3r_1[iz,]=dss004(r3l,r3u,nr,c3_1[iz,]);
  }
#
# c2 BCs
  for(iz in 1:nz){
    c2r_1[iz,1] =-(1/D2_1)*(k12f_1*c1_1[iz]-
                  k12r_1*c2_1[iz,1]);
    c2r_1[iz,nr]=-(1/D2_1)*(k23f_1*c2_1[iz,nr]-
                  k23r_1*c3_1[iz,1]);
  }
#
# c3 BCs
  for(iz in 1:nz){
    c3r_1[iz,1]=-(1/D3_1)*(k23f_1*c2_1[iz,nr]-
                  k23r_1*c3_1[iz,1]);
    c3r_1[iz,nr]=0;
  }
#
# c2rr
  nl=2;nu=2;
  c2rr_1=matrix(0,nrow=nz,ncol=nr);
  for(iz in 1:nz){
    c2rr_1[iz,]=dss044(r2l,r2u,nr,c2_1[iz,],
                      c2r_1[iz,],nl,nu);
  }
#
# c3rr
```

```
  nl=2;nu=2;
  c3rr_1=matrix(0,nrow=nz,ncol=nr);
  for(iz in 1:nz){
    c3rr_1[iz,]=dss044(r3l,r3u,nr,c3_1[iz,],
                        c3r_1[iz,],nl,nu);
  }
#
# c1t_1, c2t_1, c3t_1
  c1t_1=rep(0,nz);
  c2t_1=matrix(0,nrow=nz,ncol=nr);
  c3t_1=matrix(0,nrow=nz,ncol=nr);
  for(iz in 1:nz){
    c1t_1[iz]=-vc1_1*c1z_1[iz]-(2/r21)*
              (k12f_1*c1_1[iz]-k12r_1*c2_1[iz,1]);
  for(ir in 1:nr){
    c2t_1[iz,ir]=D2_1*(c2rr_1[iz,ir]+
                 (1/r2[ir])*c2r_1[iz,ir]);
    c3t_1[iz,ir]=D3_1*(c3rr_1[iz,ir]+(1/r3[ir])*
                 c3r_1[iz,ir])-kr_12*c3_1[iz,ir]*
                 c3_2[iz,ir];
  }
  }
  c1t_1[1]=0;
#
# One vector, two matrices to one vector
  ut=rep(0,2*(nz+2*nznr));
  for(iz in 1:nz){
    ut[iz]=c1t_1[iz];
  for(ir in 1:nr){
    izir=(iz-1)*nr+ir;
    ut[nz+izir]     =c2t_1[iz,ir];
    ut[nz+izir+nznr]=c3t_1[iz,ir];
  }
  }
```

```
#
# Second component ODE/MOL
#
# Boundary condition
  c1_2[1]=c1e_2;
#
# c1z
  c1z_2=vanl(zl,zu,nz,c1_2,vc1_2);
#
# c2r
  c2r_2=matrix(0,nrow=nz,ncol=nr);
  for(iz in 1:nz){
    c2r_2[iz,]=dss004(r2l,r2u,nr,c2_2[iz,]);
  }
#
# c3r
  c3r_2=matrix(0,nrow=nz,ncol=nr);
  for(iz in 1:nz){
    c3r_2[iz,]=dss004(r3l,r3u,nr,c3_2[iz,]);
  }
#
# c2 BCs
  for(iz in 1:nz){
    c2r_2[iz,1] =-(1/D2_2)*(k12f_2*c1_2[iz]-
                  k12r_2*c2_2[iz,1]);
    c2r_2[iz,nr]=-(1/D2_2)*(k23f_2*c2_2[iz,nr]-
                  k23r_2*c3_2[iz,1]);
  }
#
# c3 BCs
  for(iz in 1:nz){
    c3r_2[iz,1] =-(1/D3_2)*(k23f_2*c2_2[iz,nr]-
                  k23r_2*c3_2[iz,1]);
    c3r_2[iz,nr]=0;
```

```
    }
#
# c2rr
  nl=2;nu=2;
  c2rr_2=matrix(0,nrow=nz,ncol=nr);
  for(iz in 1:nz){
    c2rr_2[iz,]=dss044(r2l,r2u,nr,c2_2[iz,],
                       c2r_2[iz,],nl,nu);
  }
#
# c3rr
  nl=2;nu=2;
  c3rr_2=matrix(0,nrow=nz,ncol=nr);
  for(iz in 1:nz){
    c3rr_2[iz,]=dss044(r3l,r3u,nr,c3_2[iz,],
                       c3r_2[iz,],nl,nu);
  }
#
# c1t_2, c2t_2, c3t_2
  c1t_2=rep(0,nz);
  c2t_2=matrix(0,nrow=nz,ncol=nr);
  c3t_2=matrix(0,nrow=nz,ncol=nr);
  for(iz in 1:nz){
    c1t_2[iz]=-vc1_2*c1z_2[iz]-(2/r2l)*
              (k12f_2*c1_2[iz]-k12r_2*c2_2[iz,1]);
    for(ir in 1:nr){
    c2t_2[iz,ir]=D2_2*(c2rr_2[iz,ir]+(1/r2[ir])*
                 c2r_2[iz,ir]);
    c3t_2[iz,ir]=D3_2*(c3rr_2[iz,ir]+(1/r3[ir])*
                 c3r_2[iz,ir])-kr_12*c3_1[iz,ir]*
                 c3_2[iz,ir];
  }
  }
  c1t_2[1]=0;
```

```
#
# One vector, two matrices to one vector
  for(iz in 1:nz){
    ut[iz+nznr2]=c1t_2[iz];
    for(ir in 1:nr){
    izir=(iz-1)*nr+ir;
    ut[nz+izir+nznr2]       =c2t_2[iz,ir];
    ut[nz+izir+nznr+nznr2]=c3t_2[iz,ir];
    }
  }
#
# Increment calls to pde1a
  ncall<<-ncall+1;
#
# Return derivative vector
  return(list(c(ut)));
}
```

Listing 5.2: ODE/MOL routine for eqs. (5.1) to (5.6)

pde1a in Listing 5.2 is similar to pde1a in Listing 3.2 so the diferences will be emphasized.

- The function is defined.

```
  pde1a=function(t,u,parms){
#
# Function pde1a computes the t derivative
# vector of two components
#
# c1t_1(z,t), c2t_1(r,z,t), c3t_1(r,z,t)
#
# c1t_2(z,t), c2t_2(r,z,t), c3t_2(r,z,t)
```

u is the vector of $2(nz + 2(nz)(nr)) = 2(21 + 2(21)(6)) = 546$ MOL/ODEs.

- The first and second components are placed in one vector and two matrices.

```
#
# First component, one vector to one vector,
# two matrices
  c1_1=rep(0,nz);
  c2_1=matrix(0,nrow=nz,ncol=nr);
  c3_1=matrix(0,nrow=nz,ncol=nr);
  for(iz in 1:nz){
    c1_1[iz]=u[iz];
  for(ir in 1:nr){
    izir=(iz-1)*nr+ir;
    c2_1[iz,ir]=u[izir+nz];
    c3_1[iz,ir]=u[izir+nz+nznr];
  }
  }
#
# Second component, one vector to one vector,
# two matrices
  c1_2=rep(0,nz);
  c2_2=matrix(0,nrow=nz,ncol=nr);
  c3_2=matrix(0,nrow=nz,ncol=nr);
  for(iz in 1:nz){
    c1_2[iz]=u[iz+nznr2];
  for(ir in 1:nr){
    izir=(iz-1)*nr+ir;
    c2_2[iz,ir]=u[izir+nz+nznr2];
    c3_2[iz,ir]=u[izir+nz+nznr+nznr2];
  }
  }
```

nznr=nz*nr and nznr2=nz+2(nz*nr) are defined in the main program of Listing 5.1.

- The ODE/MOL coding for the first component is the same as in Listing 3.2.

```
#
# First component ODE/MOL
#
# Boundary condition
  c1_1[1]=c1e_1;

        .

        .

        .

#
# One vector, two matrices to one vector
  ut=rep(0,2*(nz+2*nznr));
  for(iz in 1:nz){
    ut[iz]=c1t_1[iz];
    for(ir in 1:nr){
    izir=(iz-1)*nr+ir;
    ut[nz+izir]     =c2t_1[iz,ir];
    ut[nz+izir+nznr]=c3t_1[iz,ir];
  }
  }
```

The first component ODE t derivatives are placed in ut[1] to ut[273]. The bimolecular reaction rate is included in the coding of c3t_1.

```
c3t_1[iz,ir]=D3_1*(c3rr_1[iz,ir]+(1/r3[ir])*
             c3r_1[iz,ir])-kr_12*c3_1[iz,ir]*
             c3_2[iz,ir];
```

- The coding for the second component follows directly from that for the first component with the indices displaced by 273, that is, the second component ODE t derivatives are placed in ut[274] to ut[546].

```
#
# Second component ODE/MOL
#
# Boundary condition
  c1_2[1]=c1e_2;
    .

    .

    .

#
# One vector, two matrices to one vector
  for(iz in 1:nz){
    ut[iz+nznr2]=c1t_2[iz];
    for(ir in 1:nr){
    izir=(iz-1)*nr+ir;
    ut[nz+izir+nznr2]       =c2t_2[iz,ir];
    ut[nz+izir+nznr+nznr2]=c3t_2[iz,ir];
  }
  }
```

The second component ODE t derivatives are placed in ut[274] to ut[546]. The bimolecular reaction rate is included in the coding of c3t_2.

```
c3t_2[iz,ir]=D3_2*(c3rr_2[iz,ir]+(1/r3[ir])*
             c3r_2[iz,ir])-kr_12*c3_1[iz,ir]*
             c3_2[iz,ir];
```

- The number of calls to pde1a is incremented and returned to the main program of Listing 5.1.

```
#
# Increment calls to pde1a
  ncall<<-ncall+1;
```

- The 546-vector of t derivatves is returned to lsodes as a list and a vector (from c).

```
#
# Return derivative vector
  return(list(c(ut)));
}
```

The final } concludes pde1a.

This complete the programming of eqs. (5.1) to (5.6). The output is discussed next.

(5.2.3) Model output

Abbreviated numerical output is in Table 5.2.

[1] 6

[1] 547

ncall = 757

t	z	c1_1(z,t)	c21_1(r,z,t)	c2u_1(r,z,t)
t	z	c1_1(z,t)	c31_1(r,z,t)	c3u_1(r,z,t)
0.00	0.00	1.0000	0.0000	0.0000
0.00	0.00	1.0000	0.0000	0.0000
0.00	0.10	0.0000	0.0000	0.0000
0.00	0.10	0.0000	0.0000	0.0000
0.00	0.20	0.0000	0.0000	0.0000
0.00	0.20	0.0000	0.0000	0.0000
0.00	0.30	0.0000	0.0000	0.0000
0.00	0.30	0.0000	0.0000	0.0000

```
0.00    0.40    0.0000      0.0000      0.0000
0.00    0.40    0.0000      0.0000      0.0000

0.00    0.50    0.0000      0.0000      0.0000
0.00    0.50    0.0000      0.0000      0.0000

0.00    0.60    0.0000      0.0000      0.0000
0.00    0.60    0.0000      0.0000      0.0000

0.00    0.70    0.0000      0.0000      0.0000
0.00    0.70    0.0000      0.0000      0.0000

0.00    0.80    0.0000      0.0000      0.0000
0.00    0.80    0.0000      0.0000      0.0000

0.00    0.90    0.0000      0.0000      0.0000
0.00    0.90    0.0000      0.0000      0.0000

0.00    1.00    0.0000      0.0000      0.0000
0.00    1.00    0.0000      0.0000      0.0000
                    .           .           .
                    .           .           .

        Output for t = 0.4,0.8,1.2,1.6 removed

                    .           .
                    .           .

   t       z    c1_1(z,t)   c21_1(r,z,t)   c2u_1(r,z,t)
   t       z    c1_1(z,t)   c31_1(r,z,t)   c3u_1(r,z,t)
2.00    0.00    1.0000      0.9993         0.5075
2.00    0.00    1.0000      0.5071         0.3659

2.00    0.10    0.8589      0.8582         0.4155
```

2.00	0.10	0.8589	0.4153	0.2930
2.00	0.20	0.7245	0.7239	0.3284
2.00	0.20	0.7245	0.3282	0.2234
2.00	0.30	0.5989	0.5984	0.2509
2.00	0.30	0.5989	0.2507	0.1623
2.00	0.40	0.4839	0.4834	0.1850
2.00	0.40	0.4839	0.1849	0.1120
2.00	0.50	0.3812	0.3808	0.1314
2.00	0.50	0.3812	0.1313	0.0731
2.00	0.60	0.2919	0.2916	0.0896
2.00	0.60	0.2919	0.0895	0.0448
2.00	0.70	0.2163	0.2161	0.0582
2.00	0.70	0.2163	0.0581	0.0256
2.00	0.80	0.1541	0.1539	0.0357
2.00	0.80	0.1541	0.0357	0.0135
2.00	0.90	0.1046	0.1044	0.0204
2.00	0.90	0.1046	0.0203	0.0064
2.00	1.00	0.0675	0.0674	0.0109
2.00	1.00	0.0675	0.0109	0.0028

t	z	c1_2(z,t)	c21_2(r,z,t)	c2u_2(r,z,t)
t	z	c1_2(z,t)	c31_2(r,z,t)	c3u_2(r,z,t)
0.00	0.00	1.0000	0.0000	0.0000
0.00	0.00	1.0000	0.0000	0.0000

0.00	0.10	0.0000	0.0000	0.0000
0.00	0.10	0.0000	0.0000	0.0000
0.00	0.20	0.0000	0.0000	0.0000
0.00	0.20	0.0000	0.0000	0.0000
0.00	0.30	0.0000	0.0000	0.0000
0.00	0.30	0.0000	0.0000	0.0000
0.00	0.40	0.0000	0.0000	0.0000
0.00	0.40	0.0000	0.0000	0.0000
0.00	0.50	0.0000	0.0000	0.0000
0.00	0.50	0.0000	0.0000	0.0000
0.00	0.60	0.0000	0.0000	0.0000
0.00	0.60	0.0000	0.0000	0.0000
0.00	0.70	0.0000	0.0000	0.0000
0.00	0.70	0.0000	0.0000	0.0000
0.00	0.80	0.0000	0.0000	0.0000
0.00	0.80	0.0000	0.0000	0.0000
0.00	0.90	0.0000	0.0000	0.0000
0.00	0.90	0.0000	0.0000	0.0000
0.00	1.00	0.0000	0.0000	0.0000
0.00	1.00	0.0000	0.0000	0.0000

.

.

.

Output for t = 0.4,0.8,1.2,1.6 removed

t	z	c1_2(z,t)	c21_2(r,z,t)	c2u_2(r,z,t)
t	z	c1_2(z,t)	c31_2(r,z,t)	c3u_2(r,z,t)
2.00	0.00	1.0000	0.9993	0.5075
2.00	0.00	1.0000	0.5071	0.3659
2.00	0.10	0.8589	0.8582	0.4155
2.00	0.10	0.8589	0.4153	0.2930
2.00	0.20	0.7245	0.7239	0.3284
2.00	0.20	0.7245	0.3282	0.2234
2.00	0.30	0.5989	0.5984	0.2509
2.00	0.30	0.5989	0.2507	0.1623
2.00	0.40	0.4839	0.4834	0.1850
2.00	0.40	0.4839	0.1849	0.1120
2.00	0.50	0.3812	0.3808	0.1314
2.00	0.50	0.3812	0.1313	0.0731
2.00	0.60	0.2919	0.2916	0.0896
2.00	0.60	0.2919	0.0895	0.0448
2.00	0.70	0.2163	0.2161	0.0582
2.00	0.70	0.2163	0.0581	0.0256
2.00	0.80	0.1541	0.1539	0.0357
2.00	0.80	0.1541	0.0357	0.0135
2.00	0.90	0.1046	0.1044	0.0204
2.00	0.90	0.1046	0.0203	0.0064

```
2.00    1.00    0.0675      0.0674          0.0109
2.00    1.00    0.0675      0.0109          0.0028
```

Table 5.2: Abbreviated output for eqs. (5.1) to (5.6)

We can note the following details about this output.

- The dimensions of the solution matrix from `lsodes`, `out`, are `out(6,547)` as expected.

  `[1] 6`

  `[1] 547`

- Since the equations for components one and two are the same, and the parameters defined in Listing 5.1 are the same, the solutions are the same. For example,

```
Component one

    t       z   c1_1(z,t)   c21_1(r,z,t)   c2u_1(r,z,t)
    t       z   c1_1(z,t)   c31_1(r,z,t)   c3u_1(r,z,t)
 2.00    1.00     0.0675        0.0674         0.0109
 2.00    1.00     0.0675        0.0109         0.0028

Component two

    t       z   c1_2(z,t)   c21_2(r,z,t)   c2u_2(r,z,t)
    t       z   c1_2(z,t)   c31_2(r,z,t)   c3u_2(r,z,t)
 2.00    1.00     0.0675        0.0674         0.0109
 2.00    1.00     0.0675        0.0109         0.0028
```

This test is worthwhile since different solutions would indicate a programming error.

- The computational effort is modest.

  ```
  ncall = 757
  ```

The numerical solutions are displayed in the 12 graphs produced from the main program of Listing 5.1. Component one appears in Figs. 5.1. Component two is the same so the graphical output is not included here.

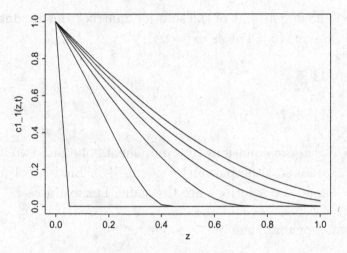

Figure 5.1a: Numerical solution $c_{1,1}(z, t)$

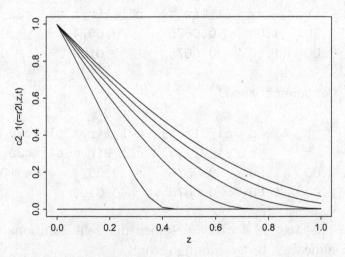

Figure 5.1b: Numerical solution $c_{2,1}(r = r_{2l}, z, t)$

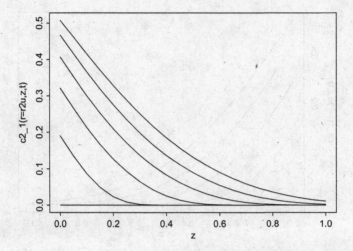

Figure 5.1c: Numerical solution $c_{2,1}(r = r_{2u}, z, t)$

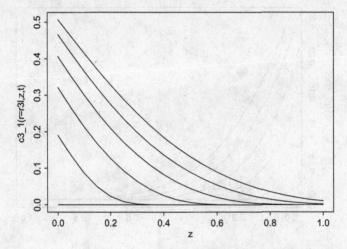

Figure 5.1d: Numerical solution $c_{3,1}(r = r_{3l}, z, t)$

The graphical output for the bimolecular reaction rate term is in Figs. 5.2.

With this verification of the solutions for the two components (the solutions are the same), differences in the components can now be studied, for example, by using different parameters.

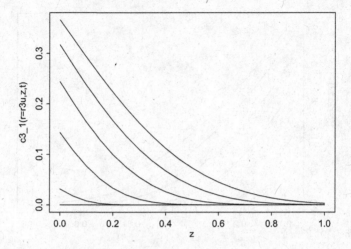

Figure 5.1e: Numerical solution $c_{3,1}(r = r_{3u}, z, t)$

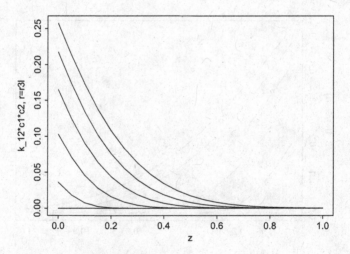

Figure 5.2a: $k_{12}c_{3,1}(r = r_{3l}, z, t)c_{3,2}(r = r_{3l}, z, t)$

This example of analyzing the bimolecular reaction rate term is intended to demonstrate how individual PDE RHS terms can be computed from the PDE solution and displayed. This can be done as well for the spatial derivatives in r and z. By adding all of the RHS terms, the LHS derivative in t can also be computed

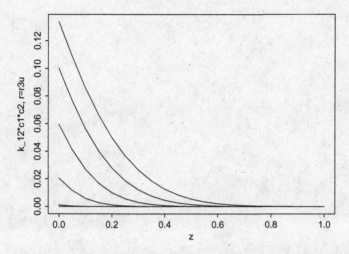

Figure 5.2b: $k_{12}c_{3,1}(r = r_{3u}, z, t)c_{3,2}(r = r_{3u}, z, t)$

and displayed. Thus, the properties of the model can be observed in full detail which can be very useful when developing a PDE application and implementation.

(5.3) Summary and Conclusions

A multicomponent three PDE BBB model is developed in this chapter. The solution of the resulting PDE system is a straightforward extension of the single component model solutions of Chapter 3. As discussed, the multicomponent extension can be applied to reaction between components in the tissue, e.g., O_2 and a nutrient, or two drugs acting simultaneously in the tissue.

Consideration is given next to the application of the preceding models with parameter values that reflect a physical BBB system.

Reference

[1] The theory of BDF integration is discussed briefly in
http://www.scholarpedia.org/article/
Backward_differentiation_formulas

Chapter 6

Application of Calibrated PDE Models to the BBB

(6.1) Introduction

In the preceding chapters, partial differential equation (PDE) models are considered with nominal values for the model parameters. In this concluding chapter, the application of these models to the dynamic analysis of the blood brain barrier (BBB) is considered with model parameters that reflect a physical description of the BBB ([1, 2]).

The first model considered is based on eqs. (3.1) to (3.4) as implemented in the main program and method of lines (MOL) ordinary differential equation (ODE) routines of Listings 3.1 and 3.2.

(6.2) One Component Model

The model of eqs. (3.1) to (3.4) is implemented in the following routines for the parameters in Table 6.1 in CGS (centimeter grams seconds) units.

parameter	definition	value with units
z_l	capillary length	1 cm
r_{2l}	inner membrane radius	25 μm[1]
r_{2u}	outer membrane radius	27 μm
r_{3l}	inner tissue radius	27 μm
r_{3u}	outer tissue radius	127 μm
v_{c1}	capillary blood velocity	0.001 cm-s^{-1}
c_{1e}	entering blood concentration	1 normalized
k_1	reaction rate constant	0 s^{-1}
D_2, D_3	membrane, tissue diffusivities, respectively	1×10^{-7} cm^2-s^{-1}
k_{f12}, k_{r12}	forward, reverse blood-membrane transfer coefficients, respectively	1×10^{-4} cm-s^{-1}
k_{f23}, k_{r23}	forward, reverse membrane-tissue transfer coefficients, respectively	1×10^{-4} cm-s^{-1}

Table 6.1: Parameters for eqs. (3.1) to (3.4)

[1] 1 μm = 1 micrometer = 1 micron = 1×10^{-4} cm.

The numerical parameters in Table 6.1 are based on reported values or estimates that give solutions to eqs. (3.1) to (3.4) which are considered realistic and reasonable. However, there is room for further refinement.

One approach to the refinement of the parameters is to make changes in the parameters so as to facilitate comparisons of the solutions for the first case (before the changes) with the second case (after the changes have been made). This might even be done so that the solutions do not change to elucidate expected dependencies on the parameters. For example, if $z_l = 1$ cm, $v_{c1} = 0.001$ cm-s^{-1} are changed to $z_l = 0.1$ cm, $v_{c1} = 0.0001$ cm-s^{-1}, the solution of eqs. (3.1) to (3.4) remains essentially unchanged. The effect of these changes is left as an exercise for the reader.

The routines for eqs. (3.1) to (3.4) with the parameters of Table 6.1 are considered next.

(6.2.1) Main program

A main program with the parameters of Table 6.1 follows.

```
#
# Calibrated three PDE BBB model
#
# Delete previous workspaces
  rm(list=ls(all=TRUE))
#
# Access ODE integrator
  library("deSolve");
#
# Access functions for numerical solution
  setwd("f:/BBB/chap6/oneComp");
  source("pde1a.R");
  source("dss004.R");
  source("dss044.R");
  source("vanl.R");
#
# Model parameters
        u10=0;           u20=0;
        u30=0; v_c1=1.0e-03;
  c1e=1.0e-00;           k1=0;
   D2=1.0e-07;    D3=1.0e-07;
 k12f=1.0e-04; k12r=1.0e-04;
 k23f=1.0e-04; k23r=1.0e-04;
#
# Initial condition
  nz=21;nr=6;
  nznr=nz*nr;
  u0=rep(0,nz+2*nznr);
```

```
    for(iz in 1:nz){
      u0[iz]=u10;
    for(ir in 1:nr){
      izir=(iz-1)*nr+ir;
      u0[nz+izir]      =u20;
      u0[nz+izir+nznr]=u30;
    }
    }
#
# Grid in z
  zl=0;zu=1;
  z=seq(from=zl,to=zu,by=(zu-zl)/(nz-1));
#
# Grid in r for c2
  r2l=0.025;r2u=0.027;
  r2=seq(from=r2l,to=r2u,by=(r2u-r2l)/(nr-1));
#
# Grid in r for c3
  r3l=0.027;r3u=0.127;
  r3=seq(from=r3l,to=r3u,by=(r3u-r3l)/(nr-1));
#
# Interval in t
  t0=0;tf=1000;nout=6;
  tout=seq(from=t0,to=tf,by=(tf-t0)/(nout-1));
  ncall=0;
#
# ODE integration
  out=lsodes(y=u0,times=tout,func=pde1a,
      sparsetype="sparseint",rtol=1e-6,
      atol=1e-6,maxord=5);
  nrow(out)
  ncol(out)
#
# Store solution
```

```
c1 =matrix(0,nrow=nz,ncol=nout);
c2l=matrix(0,nrow=nz,ncol=nout);
c2u=matrix(0,nrow=nz,ncol=nout);
c3l=matrix(0,nrow=nz,ncol=nout);
c3u=matrix(0,nrow=nz,ncol=nout);
t=rep(0,nout);
for(it in 1:nout){
for(iz in 1:nz){
   c1[iz,it]=out[it,iz+1];
  c2l[iz,it]=out[it,nz+1+(iz-1)*nr+1];
  c2u[iz,it]=out[it,nz+1+(iz-1)*nr+nr];
  c3l[iz,it]=out[it,nz+1+(iz-1)*nr+1+nznr];
  c3u[iz,it]=out[it,nz+1+(iz-1)*nr+nr+nznr];
      t[it]=out[it,1];
}
c1[1,it]=c1e;
}
#
# Display ncall
cat(sprintf("\n\n ncall = %2d",ncall));
#
# Display numerical solution
for(it in 1:nout){
cat(sprintf(
  "\n\n     t      z       c1(z,t)     c2l(r,z,t)
  c2u(r,z,t)"));
cat(sprintf(
  "\n      t      z       c1(z,t)     c3l(r,z,t)
  c3u(r,z,t)"));
izv=seq(from=1,to=nz,by=2);
for(iz in izv){
  cat(sprintf(
    "\n%7.1f%7.2f%12.2e%12.2e%12.2e",
    t[it],z[iz],c1[iz,it],c2l[iz,it],
```

```
        c2u[iz,it]));
      cat(sprintf(
        "\n%7.1f%7.2f%12.2e%12.2e%12.2e\n",
        t[it],z[iz],c1[iz,it],c31[iz,it],
        c3u[iz,it]));
  }
  }
#
# Plot numerical solutions
#
# c1(z,t)
  par(mfrow=c(1,1));
  matplot(
    x=z,y=c1,type="l",xlab="z",ylab="c1(z,t)",
    xlim=c(zl,zu),lty=1,main="",lwd=2,col="black");
#
# c2(r=r2l,z,t)
  par(mfrow=c(1,1));
  matplot(
    x=z,y=c2l,type="l",xlab="z",ylab="c2(r=r2l,z,t)",
    xlim=c(zl,zu),lty=1,main="",lwd=2,col="black");
#
# c2(r=r2u,z,t)
  par(mfrow=c(1,1));
  matplot(
    x=z,y=c2u,type="l",xlab="z",ylab="c2(r=r2u,z,t)",
    xlim=c(zl,zu),lty=1,main="",lwd=2,col="black");
#
# c3(r=rl3,z,t)
  par(mfrow=c(1,1));
  matplot(
    x=z,y=c3l,type="l",xlab="z",ylab="c3(r=r3l,z,t)",
    xlim=c(zl,zu),lty=1,main="",lwd=2,col="black");
#
```

```
# c3(r=ru3,z,t)
  par(mfrow=c(1,1));
  matplot(
    x=z,y=c3u,type="l",xlab="z",ylab="c3(r=r3u,z,t)",
    xlim=c(zl,zu),lty=1,main="",lwd=2,col="black");
```

Listing 6.1: Main program for Listing 6.1

We can note the following details about Listing 6.1.

- Previous workspaces are deleted.

```
#
# Calibrated three PDE BBB model
#
# Delete previous workspaces
  rm(list=ls(all=TRUE))
```

- The integrator library deSolve and the subordinate routines for the MOL solution are accessed.

```
#
# Access ODE integrator
  library("deSolve");
#
# Access functions for numerical solution
  setwd("f:/BBB/chap6");
  source("pde1a.R");
  source("dss004.R");
  source("dss044.R");
  source("van1.R");
```

The MOL/ODE routine is pde1a of Listing 3.2. The routines dss004, dss040, van1 are discussed after Listing 3.1.

- The parameters of Table 6.1 are programmed.

```
#
# Model parameters
        u10=0;            u20=0;
        u30=0; v_c1=1.0e-03;
  c1e=1.0e-00;            k1=0;
    D2=1.0e-07;    D3=1.0e-07;
  k12f=1.0e-04; k12r=1.0e-04;
  k23f=1.0e-04; k23r=1.0e-04;
```

- ICs $c_1(r, t = 0), c_2(r, z, t = 0), c_3(r, z, t = 0)$ (eqs. (3.2a), (3.3a), (3.4a)) are programmed.

```
#
# Initial condition
  nz=21;nr=6;
  nznr=nz*nr;
  u0=rep(0,nz+2*nznr);
  for(iz in 1:nz){
    u0[iz]=u10;
  for(ir in 1:nr){
    izir=(iz-1)*nr+ir;
    u0[nz+izir]      =u20;
    u0[nz+izir+nznr]=u30;
  }
  }
```

- An axial grid of nz=21 points is defined for $z_l = 0 \leq z \leq z_u = 1$, so that $z = 0, 0.05, \ldots, 1$ cm. The capillary therefore has a length of 1 cm or 10^4 μm.

```
#
# Grid in z
  zl=0;zu=1;
  z=seq(from=zl,to=zu,by=(zu-zl)/(nz-1));
```

- A radial grid of 6 points is defined for the membrane, $r_{2l} = 0.025 \leq r_2 \leq r_{2u} = 0.027$, so that $r_2 = 0.025, 0.0254, 0.0258, 0.0262, 0.0266, 0.027$ cm. The membrane therefore has a thickness of 0.002 cm or 20 μm. Also, the capillary has a radius of 0.025 cm or 250 μm.

```
#
# Grid in r for c2
  r2l=0.025;r2u=0.027;
  r2=seq(from=r2l,to=r2u,by=(r2u-r2l)/(nr-1));
  print("r2");
```

- A radial grid of 6 points is defined for the tissue, $r_{3l} = 0.027 \leq r_3 \leq r_{3u} = 0.127$, so that $r_3 = 0.027, 0.047, 0.067, 0.087, 0.107, 0.127$ cm. The tissue therefore has a thickness of 0.1 cm or 1000 μm.

```
#
# Grid in r for c3
  r3l=0.027;r3u=0.127;
  r3=seq(from=r3l,to=r3u,by=(r3u-r3l)/(nr-1));
```

- An interval in t of 6 points is defined for $0 \leq t \leq 1000$ so that $t = 0, 200, \ldots, 1000$ s.

```
#
# Interval in t
  t0=0;tf=1000;nout=6;
  tout=seq(from=t0,to=tf,by=(tf-t0)/(nout-1));
  ncall=0;
```

- The system of 273 MOL/ODEs is integrated by the library integrator lsodes (available in deSolve). The inputs to lsodes are the MOL/ODE function, pde1a, the IC vector u0, and the vector of output values of

t, `tout`. The length of `u0` (e.g., 273) informs `lsodes` how many ODEs are to be integrated. `func,y,times` are reserved names.

```
#
# ODE integration
  out=lsodes(y=u0,times=tout,func=pde1a,
    sparsetype="sparseint",rtol=1e-6,
    atol=1e-6,maxord=5);
  nrow(out)
  ncol(out)
```

The numerical solution to the ODEs is returned in matrix `out`. In this case, `out` has the dimensions $nout \times (nz + 2(nz)(nr) + 1) = 6 \times 21 + (2)(21)(6) + 1 = 274$, which are confirmed by the output from `nrow(out),ncol(out)` (included in the numerical output considered subsequently).

The offset $273+1 = 274$ is required since the first element of each column has the output t (also in `tout`), and the $2,\ldots,273+1 = 2,\ldots,274$ column elements have the 273 ODE solutions.

- Selected solutions of the 273 ODEs returned in `out` by `lsodes` are placed in arrays `c1, c21, c2u, c31, c3u`.

```
#
# Store solution
  c1 =matrix(0,nrow=nz,ncol=nout);
  c21=matrix(0,nrow=nz,ncol=nout);
  c2u=matrix(0,nrow=nz,ncol=nout);
  c31=matrix(0,nrow=nz,ncol=nout);
  c3u=matrix(0,nrow=nz,ncol=nout);
  t=rep(0,nout);
  for(it in 1:nout){
  for(iz in 1:nz){
```

```
  c1[iz,it]=out[it,iz+1];
  c2l[iz,it]=out[it,nz+1+(iz-1)*nr+1];
  c2u[iz,it]=out[it,nz+1+(iz-1)*nr+nr];
  c3l[iz,it]=out[it,nz+1+(iz-1)*nr+1+nznr];
  c3u[iz,it]=out[it,nz+1+(iz-1)*nr+nr+nznr];
     t[it]=out[it,1];
}
c1[1,it]=c1e;
}
```

Again, the offsets `iz+1`, `nz+1` are required since the first element of each column of `out` has the value of t. $c_1(z = 0, t) =$ `c1[1,it]=c1e` sets BC (3.2b) since this value is not returned by `lsodes` (it is defined algebraically in `pde1a` and not by the integration of an ODE at $z = 0$). $c_2(r = r_{2l}, z, t), c_2(r = r_{2u}, z, t)$ are placed in `c2l`, `c2u`. $c_3(r = r_{3l}, z, t), c_3(r = r_{3u}, z, t)$ are placed in `c3l`, `c3u`.

- The number of calls to `pde1a` is displayed at the end of the solution.

```
#
# Display ncall
  cat(sprintf("\n\n ncall = %2d",ncall));
```

- Five selected values of the solution are plotted against z and parametrically in t.

```
#
# Plot numerical solutions
#
# c1(z,t)
  par(mfrow=c(1,1));
  matplot(
    x=z,y=c1,type="l",xlab="z",
    ylab="c1(z,t)",xlim=c(zl,zu),
    lty=1,main="",lwd=2,col="black");
```

```
#
# c2(r=r2l,z,t)
  par(mfrow=c(1,1));
  matplot(
    x=z,y=c2l,type="l",xlab="z",
    ylab="c2(r=r2l,z,t)",xlim=c(zl,zu),
    lty=1,main="",lwd=2,col="black");
#
# c2(r=r2u,z,t)
  par(mfrow=c(1,1));
  matplot(
    x=z,y=c2u,type="l",xlab="z",
    ylab="c2(r=r2u,z,t)"xlim=c(zl,zu),
    lty=1,main="",lwd=2,col="black");
#
# c3(r=rl3,z,t)
  par(mfrow=c(1,1));
  matplot(
    x=z,y=c3l,type="l",xlab="z",
    ylab="c3(r=r3l,z,t)",xlim=c(zl,zu),
    lty=1,main="",lwd=2,col="black");
#
# c3(r=ru3,z,t)
  par(mfrow=c(1,1));
  matplot(
    x=z,y=c3u,type="l",xlab="z",
    ylab="c3(r=r3u,z,t)",xlim=c(zl,zu),
    lty=1,main="",lwd=2,col="black");
```

This completes the main program.

(6.2.2) ODE/MOL routine

The ODE/MOL routine pde1a is the same as in Listing 3.2 and therefore is not listed here.

(6.2.3) Model output

Abbreviated output from the main program of Listing 6.1 and the MOL/ODE routine pde1a of Listing 3.2 follows.

[1] 6

[1] 274

ncall = 970

t	z	c1(z,t)	c2l(r,z,t)	c2u(r,z,t)
t	z	c1(z,t)	c3l(r,z,t)	c3u(r,z,t)
0.0	0.00	1.00e+00	0.00e+00	0.00e+00
0.0	0.00	1.00e+00	0.00e+00	0.00e+00
0.0	0.10	0.00e+00	0.00e+00	0.00e+00
0.0	0.10	0.00e+00	0.00e+00	0.00e+00
0.0	0.20	0.00e+00	0.00e+00	0.00e+00
0.0	0.20	0.00e+00	0.00e+00	0.00e+00
0.0	0.30	0.00e+00	0.00e+00	0.00e+00
0.0	0.30	0.00e+00	0.00e+00	0.00e+00
0.0	0.40	0.00e+00	0.00e+00	0.00e+00
0.0	0.40	0.00e+00	0.00e+00	0.00e+00
0.0	0.50	0.00e+00	0.00e+00	0.00e+00
0.0	0.50	0.00e+00	0.00e+00	0.00e+00
0.0	0.60	0.00e+00	0.00e+00	0.00e+00

0.0	0.60	0.00e+00	0.00e+00	0.00e+00
0.0	0.70	0.00e+00	0.00e+00	0.00e+00
0.0	0.70	0.00e+00	0.00e+00	0.00e+00
0.0	0.80	0.00e+00	0.00e+00	0.00e+00
0.0	0.80	0.00e+00	0.00e+00	0.00e+00
0.0	0.90	0.00e+00	0.00e+00	0.00e+00
0.0	0.90	0.00e+00	0.00e+00	0.00e+00
0.0	1.00	0.00e+00	0.00e+00	0.00e+00
0.0	1.00	0.00e+00	0.00e+00	0.00e+00

.
.
.

Output for t = 200,400,600,900 removed

.
.
.

t	z	$c1(z,t)$	$c21(r,z,t)$	$c2u(r,z,t)$
t	z	$c1(z,t)$	$c31(r,z,t)$	$c3u(r,z,t)$
1000.0	0.00	1.00e+00	9.38e-01	8.18e-01
1000.0	0.00	1.00e+00	7.61e-01	3.50e-03
1000.0	0.10	9.48e-01	8.85e-01	7.63e-01
1000.0	0.10	9.48e-01	7.06e-01	2.63e-03
1000.0	0.20	8.93e-01	8.27e-01	7.02e-01
1000.0	0.20	8.93e-01	6.44e-01	1.87e-03
1000.0	0.30	8.30e-01	7.61e-01	6.32e-01
1000.0	0.30	8.30e-01	5.72e-01	1.25e-03

1000.0	0.40	7.56e-01	6.83e-01	5.49e-01
1000.0	0.40	7.56e-01	4.88e-01	7.62e-04
1000.0	0.50	6.69e-01	5.93e-01	4.55e-01
1000.0	0.50	6.69e-01	3.94e-01	4.17e-04
1000.0	0.60	5.65e-01	4.88e-01	3.52e-01
1000.0	0.60	5.65e-01	2.92e-01	1.95e-04
1000.0	0.70	4.45e-01	3.72e-01	2.45e-01
1000.0	0.70	4.45e-01	1.91e-01	7.29e-05
1000.0	0.80	3.13e-01	2.51e-01	1.45e-01
1000.0	0.80	3.13e-01	1.01e-01	1.89e-05
1000.0	0.90	1.76e-01	1.32e-01	6.27e-02
1000.0	0.90	1.76e-01	3.56e-02	2.45e-06
1000.0	1.00	6.12e-02	4.24e-02	1.56e-02
1000.0	1.00	6.12e-02	6.42e-03	1.59e-07

Table 6.2: Solution for parameters in Table 6.1

We can note the following details about the solution in Table 6.2.

- The dimensions of the solution matrix from out are $6 \times 273 + 1 = 274$. The offset $+1$ includes the value of t as the first element of each of the 6 solution vectors in out.

[1] 6

[1] 274

- The homogeneous ICs for eqs. (3.1) and the BC $c_1(z = 0, t) = c_{1e} = 1$ are confirmed.
- The solution at $t = 1000$ indicates the propagation of the unit step in $c_1(z, t)$ from the BC at $z = 0$. The time scale $0 \leq t \leq 1000$ corresponds to $v_{c1} = 0.001, z_l = 1$ (1000 s are required for the step to move the length of the capillary).
- The computational effort is modest, `ncall = 970`.

The solution in Figs. 6.1 corresponds to the numerical solution in Table 6.2.

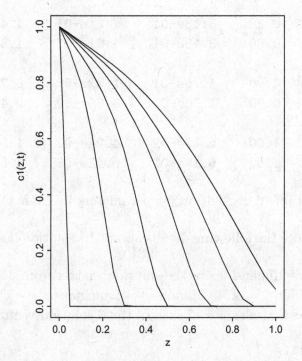

Figure 6.1a: Numerical solution for $c_1(z, t)$

Fig. 6.1a indicates the propagation of the unit step along the capillary.

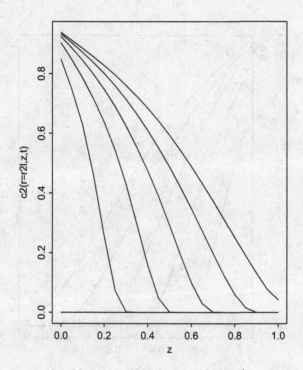

Figure 6.1b: Numerical solution for $c_2(r = r_{2l}, z, t)$

Fig. 6.1b indicates the response of the inner membrane concentration $c_2(r = r_{2l}, z, t)$ to the changing blood concentration $c_1(z, t)$ of Fig. 6.1a.

The mass transport flux at the inner membrane boundary can be computed as $(k_{12f}c_1(z, t) - k_{12r}c_2(r = r_{2l}, z, t))$ (from BC 3.3b). Note that this flux is in the positive direction of r. This additional analysis is an example of computing and displaying terms in the PDE model. It is left as an excerise for the reader.

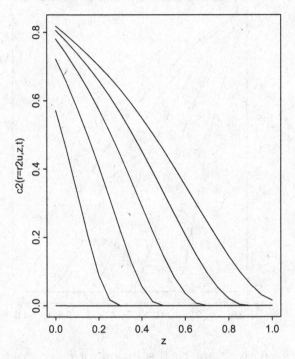

Figure 6.1c: Numerical solution for $c_2(r = r_{2u}, z = 0.5, t)$

Fig. 6.1c indicates the response of the outer membrane concentration $c_2(r = r_{2u}, z, t)$ to the inner membrane concentration $c_2(r = r_{2l}, z, t)$ of Fig. 6.1b. The decrease of the concentration $c_2(r, z, t)$ across the membrane can be observed from the solution in Table 6.2. For example, from Table 6.2, the difference

$c_{2u}(r = r_{2u} = 0.027, z = 0.5, t = 1000) - c_{2l}(r = r_{2l} = 0.025, z = 0.5, t = 1000) = $ 5.93e-01 - 4.55e-01 = 0.138.

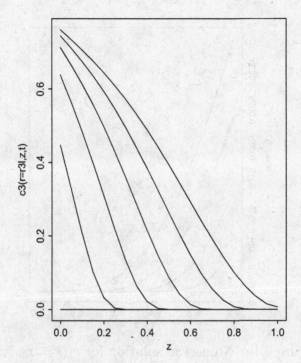

Figure 6.1d: Numerical solution for $c_3(r = r_{3l}, z, t)$

Fig. 6.1d indicates the response of the inner tissue concentration $c_3(r = r_{3l}, z, t)$ to the outer membrane concentration $c_2(r = r_{2u}, z, t)$ of Fig. 6.1c.

The mass transport flux at the inner tissue boundary can be computed as $(k_{23f}c_2(r = r_{2u}, z, t) - k_{23r}c_3(r = r_{3l}, z, t))$ (from BC 3.4b). Note that this flux is in the positive direction of r.

This additional analysis is an example of computing and displaying terms in the PDE model. It is left as an excerise for the reader.

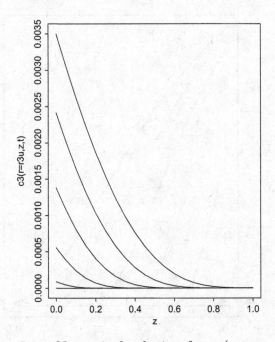

Figure 6.1e: Numerical solution for $c_3(r = r_{3u}, z, t)$

Fig. 6.1e indicates the response of the outer tissue concentration $c_3(r = r_{3u}, z, t)$ to the inner tissue concentration $c_3(r = r_{3l}, z, t)$ of Fig. 6.1d. The decrease of the concentration $c_3(r, z, t)$ across the tissue can be observed from the solution in Table 6.2. For example, from Table 6.2, the difference $c_{3u}(r = r_{3u} = 0.127, z = 0.5, t = 1000) - c_{3l}(r = r_{3u} = 0.027, z = 0.5, t = 1000) = $ 3.94e-01 - 4.17e-04 = 0.394. Also, the outer tissue concentration is small, $c_{3u}(r = r_{3u} = 0.127, z = 0.5, t = 1000) = $ 4.17e-04, indicating the transported component has not fully penetrated the tissue by $t = 1000$.

The concentration drop in the tissue is greater than in the membrane since the membrane is thinner (from Table 6.1, $0.027 - 0.025 = 0.002$ cm for the membrane and $0.127 - 0.027 = 0.1$ cm for the tissue, with numerically equal mass transfer coefficients and diffusivities for the two sections).

Fig. 6.1a indicates limited non-smoothness which will be most pronounced for $c_1(z,t)$ since eq. (3.1a) is hyperbolic (first order in z, t) while eqs. (3.1b,c) are parabolic (first order in t second order in r). In other words, eq. (3.1a) will propagate the step input in z from $c_{1e} = 1$ while eqs. (3.1b,c) will smooth the changes in r.

To reduce the gridding effect of Fig. 6.1a, the number of points in z is increased from nz=21 to nz=41. The numerical solution then changes as in Table 6.3.

[1] 6

[1] 534

ncall = 1895

t	z	c1(z,t)	c21(r,z,t)	c2u(r,z,t)
t	z	c1(z,t)	c31(r,z,t)	c3u(r,z,t)
0.0	0.00	1.00e+00	0.00e+00	0.00e+00
0.0	0.00	1.00e+00	0.00e+00	0.00e+00
0.0	0.05	0.00e+00	0.00e+00	0.00e+00
0.0	0.05	0.00e+00	0.00e+00	0.00e+00
0.0	0.10	0.00e+00	0.00e+00	0.00e+00
0.0	0.10	0.00e+00	0.00e+00	0.00e+00

.

```
              Output for z = 0.15 to 0.85 removed
                      .
                      .
                      .
   0.0    0.90    0.00e+00     0.00e+00     0.00e+00
   0.0    0.90    0.00e+00     0.00e+00     0.00e+00

   0.0    0.95    0.00e+00     0.00e+00     0.00e+00
   0.0    0.95    0.00e+00     0.00e+00     0.00e+00

   0.0    1.00    0.00e+00     0.00e+00     0.00e+00
   0.0    1.00    0.00e+00     0.00e+00     0.00e+00
                      .
                      .
                      .
              Output for t = 200 to 900 removed
                      .
                      .
                      .
     t      z     c1(z,t)     c2l(r,z,t)   c2u(r,z,t)
     t      z     c1(z,t)     c3l(r,z,t)   c3u(r,z,t)
  1000.0  0.00   1.00e+00     9.38e-01     8.18e-01
  1000.0  0.00   1.00e+00     7.61e-01     3.50e-03

  1000.0  0.05   9.75e-01     9.12e-01     7.92e-01
  1000.0  0.05   9.75e-01     7.35e-01     3.05e-03

  1000.0  0.10   9.49e-01     8.85e-01     7.64e-01
  1000.0  0.10   9.49e-01     7.07e-01     2.62e-03
                      .
                      .
                      .
```

Output for z = 0.15 to 0.45 removed

```
1000.0    0.50    6.71e-01    5.94e-01    4.56e-01
1000.0    0.50    6.71e-01    3.94e-01    3.97e-04
```

Output for z = 0.55 to 0.85 removed

```
1000.0    0.90    1.73e-01    1.27e-01    5.43e-02
1000.0    0.90    1.73e-01    2.66e-02    8.87e-07

1000.0    0.95    9.70e-02    6.56e-02    2.16e-02
1000.0    0.95    9.70e-02    7.02e-03    6.76e-08

1000.0    1.00    2.89e-02    1.71e-02    3.85e-03
1000.0    1.00    2.89e-02    7.37e-04    2.30e-09
```

Table 6.3: Solution for parameters in Table 6.1, nz=41

The only change in the main program of Listing 6.1 is

```
#
# Initial condition
  nz=41;nr=6;
```

for Table 6.3 in place of

```
#
# Initial condition
  nz=21;nr=6;
```

for Table 6.2.

The solution in Table 6.3 has an interval in z of 0.05 corresponding to nz=41 rather than the solution in Table 6.2 with an interval of 0.1. The solutions for nz=21,41 agree to approximately three figures.

Table 6.2 (nz=21)

t	z	c1(z,t)	c21(r,z,t)	c2u(r,z,t)
t	z	c1(z,t)	c31(r,z,t)	c3u(r,z,t)
1000.0	0.00	1.00e+00	9.38e-01	8.18e-01
1000.0	0.00	1.00e+00	7.61e-01	3.50e-03
1000.0	0.50	6.69e-01	5.93e-01	4.55e-01
1000.0	0.50	6.69e-01	3.94e-01	4.17e-04
1000.0	1.00	6.12e-02	4.24e-02	1.56e-02
1000.0	1.00	6.12e-02	6.42e-03	1.59e-07

Table 6.3 (nz=41)

t	z	c1(z,t)	c21(r,z,t)	c2u(r,z,t)
t	z	c1(z,t)	c31(r,z,t)	c3u(r,z,t)
1000.0	0.00	1.00e+00	9.38e-01	8.18e-01
1000.0	0.00	1.00e+00	7.61e-01	3.50e-03
1000.0	0.50	6.71e-01	5.94e-01	4.56e-01
1000.0	0.50	6.71e-01	3.94e-01	3.97e-04
1000.0	1.00	2.89e-02	1.71e-02	3.85e-03
1000.0	1.00	2.89e-02	7.37e-04	2.30e-09

Fig. 6.2a indicates the propagation of the unit step along the capillary is smoother with nz=41 than with nz=21 in Fig. 6.1a.

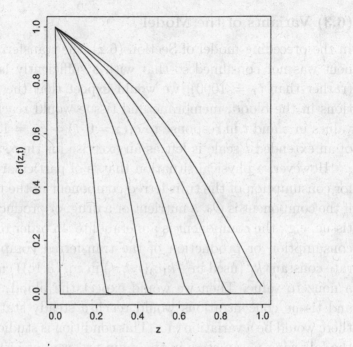

Figure 6.2a: Numerical solution for $c_1(z,t)$, nz=41

The preceding change in nz is an example of h-refinement for assessing the accuracy of a numerical solution (by comparing the solutions for two or more numbers of grid points). The conclusion in this case is that nz=21 is adequate (for three figure accuracy).

Also, the solution matrix out has dimensions $6 \times 41 + 2(41)(6) = 533$. So by increasing nz, the number of MOL/ODEs increased from 273 to 533, and generally, increasing the number of grid points must be done with some care. In the present case, ncall=1895 for nz=41 (against ncall=970 for nz=21). The increase in ncall is probably due to an increase in the stiffness of the MOL/ODEs. Additionally, the calculations within each call increased substantially because of the increase in the number of ODEs. Generally, the total number of ODEs is $O(n_z n_r)$ for this 2D PDE model.

(6.3) Variants of the Model

In the preceding model of Section (6.2), the transferred component was not consumed so that with a sufficiently large final t (rather than $t_f = 1000$), we would expect that the concentrations in the blood, membrane and tissue would reach constant values in z and t in response to $c_1(z = 0, t) = c_{1e} = 1$. This case of an extended t scale is left as an excerise for the reader.

However, a physical situation that is of particular interest is for consumption of the transferred component in the tissue, e.g., if the component is O_2, a nutrient or a drug, or production in the tissue, e.g., the component is a metabolite. In order to represent consumption or production of the transferred component, the rate constant k_1 (used in $-k_1 c_3(r, z, t)$ in eq. (3.1c)) can be given a nonzero value. Then we would expect the blood, membrane and tissue concentrations would reach a steady state in t and there would be a variation in z. This condition is studied by using the following parameters in the main program of Listing 6.1.

$$k_1 = 10$$
$$t_f = 2000$$
$$\text{nout} = 11$$
$$\text{nz} = 41$$

Table 6.4: Modified parameters

A reaction now occurs with $k_1 = 10$ and a steady state is approached with $t_f = 2000$. The number of output points is increased from 6 to 11 to accommodate the increased t_f, and **nz=41** points in z gives adequate spatial resolution.

Abbreviated numerical output for the parameters of Table 6.4 follows.

[1] 11

[1] 534

```
ncall = 2334
```

t	z	c1(z,t)	c2l(r,z,t)	c2u(r,z,t)
t	z	c1(z,t)	c3l(r,z,t)	c3u(r,z,t)
0.0	0.00	1.00e+00	0.00e+00	0.00e+00
0.0	0.00	1.00e+00	0.00e+00	0.00e+00
0.0	0.05	0.00e+00	0.00e+00	0.00e+00
0.0	0.05	0.00e+00	0.00e+00	0.00e+00
0.0	0.10	0.00e+00	0.00e+00	0.00e+00
0.0	0.10	0.00e+00	0.00e+00	0.00e+00

.
.
.

Output for z = 0.15,...,0.85 removed

.
.
.

0.0	0.90	0.00e+00	0.00e+00	0.00e+00
0.0	0.90	0.00e+00	0.00e+00	0.00e+00
0.0	0.95	0.00e+00	0.00e+00	0.00e+00
0.0	0.95	0.00e+00	0.00e+00	0.00e+00
0.0	1.00	0.00e+00	0.00e+00	0.00e+00
0.0	1.00	0.00e+00	0.00e+00	0.00e+00

.
.
.

Output for t = 200,...,800 removed

.
.
.

t	z	c1(z,t)	c2l(r,z,t)	c2u(r,z,t)
t	z	c1(z,t)	c3l(r,z,t)	c3u(r,z,t)
1000.0	0.00	1.00e+00	7.40e-01	2.41e-01
1000.0	0.00	1.00e+00	4.12e-04	1.35e-13
1000.0	0.05	9.03e-01	6.68e-01	2.17e-01
1000.0	0.05	9.03e-01	3.72e-04	1.22e-13
1000.0	0.10	8.14e-01	6.03e-01	1.96e-01
1000.0	0.10	8.14e-01	3.35e-04	1.10e-13

$\cdot$
$\cdot$
$\cdot$

Output for z = 0.15,...,0.40 removed

$\cdot$
$\cdot$
$\cdot$

1000.0	0.45	3.93e-01	2.91e-01	9.47e-02
1000.0	0.45	3.93e-01	1.62e-04	5.30e-14
1000.0	0.50	3.55e-01	2.62e-01	8.54e-02
1000.0	0.50	3.55e-01	1.46e-04	4.78e-14
1000.0	0.55	3.20e-01	2.37e-01	7.69e-02
1000.0	0.55	3.20e-01	1.32e-04	4.31e-14

$\cdot$
$\cdot$
$\cdot$

Output for z = 0.60,...,0.85 removed

$\cdot$
$\cdot$
$\cdot$

1000.0	0.90	1.35e-01	9.62e-02	2.95e-02

1000.0	0.90	1.35e-01	5.03e-05	1.65e-14
1000.0	0.95	7.87e-02	5.11e-02	1.29e-02
1000.0	0.95	7.87e-02	2.20e-05	7.16e-15
1000.0	1.00	1.59e-02	8.63e-03	1.39e-03
1000.0	1.00	1.59e-02	2.36e-06	7.61e-16

Output for t = 1200,...,1800 removed

2000.0	0.00	1.00e+00	7.40e-01	2.41e-01
2000.0	0.00	1.00e+00	4.12e-04	1.35e-13
2000.0	0.05	9.03e-01	6.68e-01	2.17e-01
2000.0	0.05	9.03e-01	3.72e-04	1.22e-13
2000.0	0.10	8.14e-01	6.03e-01	1.96e-01
2000.0	0.10	8.14e-01	3.35e-04	1.10e-13

Output for z = 0.15,...,0.40 removed

2000.0	0.45	3.93e-01	2.91e-01	9.47e-02
2000.0	0.45	3.93e-01	1.62e-04	5.30e-14
2000.0	0.50	3.55e-01	2.62e-01	8.54e-02
2000.0	0.50	3.55e-01	1.46e-04	4.78e-14

2000.0	0.55	3.20e-01	2.37e-01	7.69e-02
2000.0	0.55	3.20e-01	1.32e-04	4.31e-14
	·	·		·
	·	·		·
	·	·		·

Output for z = 0.60,...,0.85 removed

	·			·
	·			·
	·			·
2000.0	0.90	1.54e-01	1.14e-01	3.72e-02
2000.0	0.90	1.54e-01	6.36e-05	2.08e-14
2000.0	0.95	1.39e-01	1.03e-01	3.35e-02
2000.0	0.95	1.39e-01	5.73e-05	1.88e-14
2000.0	1.00	1.26e-01	9.30e-02	3.02e-02
2000.0	1.00	1.26e-01	5.17e-05	1.69e-14

Table 6.5: Solution for parameters in Table 6.1, nz=41,
$k_1 = 10$, $t_f = 2000$

We can note the following details about this output.

- The solution array out from lsodes has the dimensions out(11,41+(2*41*6)+1=534). As indicated previously, the offset +1 accommodates the value of t as the first element in each of the 11 533-vectors.

  [1] 11

  [1] 534

- The homogeneous ICs $c_1(z, t = 0) = c_2(r, z, t = 0) = c_3(r, z, t = 0) = 0$ and the BC $c_1(z = 0, t) = c_{1e} = 1$ are confirmed

- At $t = 1000, 2000$, $z = 0$, the solution has reached a steady state (independent of t).

```
1000.0   0.00   1.00e+00   7.40e-01   2.41e-01
1000.0   0.00   1.00e+00   4.12e-04   1.35e-13

2000.0   0.00   1.00e+00   7.40e-01   2.41e-01
2000.0   0.00   1.00e+00   4.12e-04   1.35e-13
```

- Similarly, at $z = 0.5$ the solution has reached a steady state.

```
1000.0   0.50   3.55e-01   2.62e-01   8.54e-02
1000.0   0.50   3.55e-01   1.46e-04   4.78e-14

2000.0   0.50   3.55e-01   2.62e-01   8.54e-02
2000.0   0.50   3.55e-01   1.46e-04   4.78e-14
```

- At $z = 1$ the solution is still evolving in t.

```
1000.0   1.00   1.59e-02   8.63e-03   1.39e-03
1000.0   1.00   1.59e-02   2.36e-06   7.61e-16

2000.0   1.00   1.26e-01   9.30e-02   3.02e-02
2000.0   1.00   1.26e-01   5.17e-05   1.69e-14
```

The solution at $t = 1800$, $z = 1$ is (not included in Table 6.5) indicates a steady state has been reached by $t = 2000$.

```
1800.0   1.00   1.26e-01   9.30e-02   3.02e-02
1800.0   1.00   1.26e-01   5.17e-05   1.69e-14

2000.0   1.00   1.26e-0    9.30e-02   3.02e-02
2000.0   1.00   1.26e-01   5.17e-05   1.69e-14
```

,

- The computational effort is acceptable, `ncall = 2334`.

These properties are reflected in the graphical output of Figs. 6.3 (Figs. 6.1 are the analogous figures for $k_1 = 0$).

The approach to a steady state is clear in all of these figures. Also, the consumption term $-k_1 c_3(r, z, t)$ (eq. (3.1c)) substantially reduces $c_3(r = r_{3l}, z, t)$ and $c_3(r = r_{3u}, z, t)$ as indicated by comparing Figs. 6.1d, 6.3d, and Figs. 6.1e, 6.3e. For example, the model can be used to elucidate the consumption of O_2, a nutrient or a therapeutic drug in the tissue. The steady state profiles will be determined by the permeability of the membrane according to the values of $k_{12f}, k_{12r}, k_{23f}, k_{23r}$ (in BCs (3.3b), (3.3c), (3.4b)) and the diffusivity of the membrane, D_3 (in eq. (3.1c)), as well as the consumption term $-k_1 c_3(r, z, t)$. In particular, the tissue concentration of a therapeutic drug might remain low if the membrane permeability is low, or the tissue concentration of a harmful substance might be high if the membrane permeability is large for that substance.

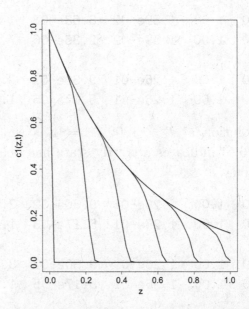

Figure 6.3a: Numerical solution for $c_1(z, t)$, $k_1 = 10$, $t_f = 2000$, nz=41

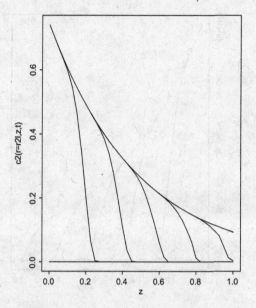

Figure 6.3b: Numerical solution for $c_2(r = r_{2l}, z, t)$, $k_1 = 10$, $t_f = 2000$, nz=41

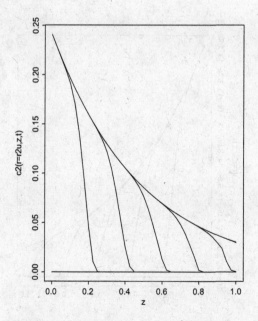

Figure 6.3c: Numerical solution for $c_2(r = r_{2u}, z, t)$, $k_1 = 10$, $t_f = 2000$, nz=41

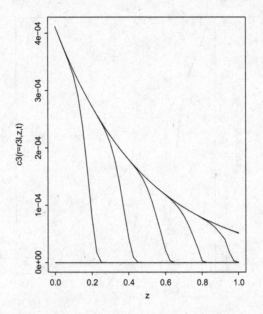

Figure 6.3d: Numerical solution for $c_3(r = r_{3l}, z, t)$, $k_1 = 10$, $t_f = 2000$, nz=41

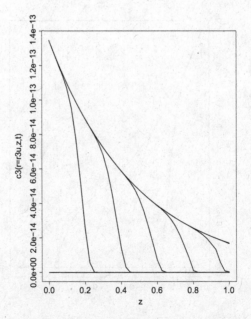

Figure 6.3e: Numerical solution for $c_3(r = r_{3u}, z, t)$, $k_1 = 10$, $t_f = 2000$, nz=41

(6.4) Summary and Conclusions

The preceding examples demonstrate the application of the one component model of Chapter 3 with parameters that numerically reflect BBB properties. The calculated profiles from eqs. (3.1) to (3.4) illustrate how the model can be used to investigate the spatiotemporal profiles of various substances that might be of interest such as O_2, a nutrient, a therapeutic drug, a harmful chemical.

These substances can be analyzed in various combinations by using multi component models as illustrated in Chapter 5. For example,

- O_2 + nutrient $\Rightarrow$ two component model
- O_2 + nutrient + metabolite $\Rightarrow$ three component model

Also, the estimation of parameters from recently reported experimental studies is an important enhancement of the model. The intention in the development of the PDE models is to hopefully provide a computer-based methodology for the quantitative analysis of the BBB system.

References

[1] Ballabh, P., A. Braun and M. Nedergaard (2004), The blood-brain barrier: an overview, Structure, regulation, and clinical implications, *Neurobiology of Disease*, **16**, pp 1–13.

[2] Pardridge, W. (2012), Drug transport across the blood-brain barrier, *Journal of Cerebral Blood Flow & Metabolism*, **32**, pp 1959–1972.

Appendix A1: Function dss004

A listing of function dss004 follows.

```
dss004=function(xl,xu;n,u) {
#
# An extensive set of documentation comments
# detailing the derivation of the following
# fourth order finite differences (FDs) is
# not given here to conserve space.  The
# derivation is detailed in Schiesser, W. E.,
# The Numerical Method of Lines Integration
# of Partial Differential Equations, Academic
# Press, San Diego, 1991.
#
# Preallocate arrays
  ux=rep(0,n);
#
# Grid spacing
  dx=(xu-xl)/(n-1);
#
# 1/(12*dx) for subsequent use
  r12dx=1/(12*dx);
#
# ux vector
#
```

```
# Boundaries (x=xl,x=xu)
  ux[1]=r12dx*(-25*u[1]+48*u[  2]-36*u[  3]+
             16*u[  4] -3*u[  5]);
  ux[n]=r12dx*(  25*u[n]-48*u[n-1]+36*u[n-2]-
             16*u[n-3 ] +3*u[n-4]);
#
# dx in from boundaries (x=xl+dx,x=xu-dx)
  ux[  2]=r12dx*(-3*u[1]-10*u[  2]+18*u[  3]-
             6*u[  4]    +u[  5]);
  ux[n-1]=r12dx*(  3*u[n]+10*u[n-1]-18*u[n-2]+
             6*u[n-3]   -u[n-4]);
#
# Interior points (x=xl+2*dx,...,x=xu-2*dx)
  for(i in 3:(n-2)){
    ux[i]=r12dx*(-u[i+2]+8*u[i+1]-
             8*u[i-1]   +u[i-2]);}
#
# All points concluded (x=xl,...,x=xu)
  return(c(ux));
}
```

The input arguments are

xl	lower boundary value of x
xu	upper boundary value of x
n	number of points in the grid in x, including the end points
u	dependent variable to be differentiated, an n-vector

The output, ux, is an n-vector of numerical values of the first derivative of u.

The finite difference (FD) approximations are a weighted sum of the dependent variable values. For example, at point i

```
#
# Interior points (x=xl+2*dx,...,x=xu-2*dx)
  for(i in 3:(n-2)){
    ux[i]=r12dx*(-u[i+2]+8*u[i+1]-
              8*u[i-1]  +u[i-2]);}
```

The weighting coefficients are -1, 8, 0, -8, 1 at points i-2, i-1, i, i+1, i+2, respectively. These weighting coefficients are antisymmetric (opposite sign) around the center point i because the computed first derivative is of odd order. If the derivative is of even order, the weighting coefficients would be symmetric (same sign) around the center point.

For i=1. the dependent variable at points i=1,2,3,4,5 is used in the FD approximation for ux[1] to remain within the x domain (fictitious points outside the x domain are not used).

```
ux[1]=r12dx*(-25*u[1]+48*u[  2]-36*u[  3]+
             16*u[  4] -3*u[  5]);
```

Similarly, for i=2, points i=1,2,3,4,5 are used in the FD approximation for ux[2] to remain within the x domain (fictitious points outside the x domain are avoided).

```
ux[  2]=r12dx*(-3*u[1]-10*u[  2]+18*u[  3]-
              6*u[  4]   +u[  5]);
```

At the right boundary $x = x_u$, points at i=n,n-1,n-2,n-3,n-4 are used for ux[n],ux[n-1] to avoid points outside the x domain.

In all cases, the FD approximations are fourth order correct in x.

Appendix A2: Function dss044

A listing of function dss044 follows.

```
dss044=function(xl,xu,n,u,ux,nl,nu) {
#
# The derivation of the finite difference
# approximations for a second derivative are
# in Schiesser, W. E., The Numerical Method
# of Lines Integration of Partial Differential
# Equations, Academic Press, San Diego, 1991.
#
# Preallocate arrays
  uxx=rep(0,n);
#
# Grid spacing
  dx=(xu-xl)/(n-1);
#
# 1/(12*dx**2) for subsequent use
  r12dxs=1/(12*dx^2);
#
# uxx vector
#
# Boundaries (x=xl,x=xu)
  if(nl==1)
    uxx[1]=r12dxs*
```

```
              (45*u[  1]-154*u[  2]+214*u[  3]-
              156*u[  4] +61*u[  5] -10*u[  6]);
  if(nu==1)
    uxx[n]=r12dxs*
              (45*u[  n]-154*u[n-1]+214*u[n-2]-
              156*u[n-3] +61*u[n-4] -10*u[n-5]);
  if(nl==2)
    uxx[1]=r12dxs*
              (-415/6*u[  1] +96*u[  2]-36*u[  3]+
              32/3*u[  4]-3/2*u[  5]-50*ux[1]*dx);
  if(nu==2)
    uxx[n]=r12dxs*
              (-415/6*u[  n] +96*u[n-1]-36*u[n-2]+
              32/3*u[n-3]-3/2*u[n-4]+50*ux[n]*dx);
#
# dx in from boundaries (x=xl+dx,x=xu-dx)
    uxx[  2]=r12dxs*
              (10*u[  1]-15*u[  2]-4*u[  3]+
              14*u[  4]- 6*u[  5]  +u[  6]);
    uxx[n-1]=r12dxs*
              (10*u[  n]-15*u[n-1]-4*u[n-2]+
              14*u[n-3]- 6*u[n-4]  +u[n-5]);
#
# Remaining interior points (x=xl+2*dx,...,
# x=xu-2*dx)
  for(i in 3:(n-2))
    uxx[i]=r12dxs*
              (-u[i-2]+16*u[i-1]-30*u[i]+
              16*u[i+1]   -u[i+2]);
#
# All points concluded (x=xl,...,x=xu)
  return(c(uxx));
}
```

The input arguments are

xl   lower boundary value of x

xu   upper boundary value of x

n   number of points in the grid in x, including the end points

u   dependent variable to be differentiated, an n-vector

ux   first derivative of u with boundary condition values, an n-vector

nl   type of boundary condition at x=xl
1: Dirichlet BC
2: Neumann BC

nu   type of boundary condition at x=xu
1: Dirichlet BC
2: Neumann BC

The output, uxx, is an n-vector of numerical values of the second derivative of u.

The finite difference (FD) approximations are a weighted sum of the dependent variable values. For example, at point i

```
for(i in 3:(n-2))
  uxx[i]=r12dxs*
         (-u[i-2]+16*u[i-1]-30*u[i]+
         16*u[i+1]   -u[i+2]);
```

The weighting coefficients are -1, 16, -30, 16, -1 at points i-2, i-1, i, i+1, i+2, respectively. These weighting coefficients are symmetric around the center point i because the computed second derivative is of even order. If the derivative is of odd order, the weighting coefficients would be antisymmetric (opposite sign) around the center point.

For nl=2 and/or nu=2 the boundary values of the first derivative are included in the FD approximation for the second derivative, uxx. For example, at x=xl (with nl=2),

```
if(nl==2)
  uxx[1]=r12dxs*
         (-415/6*u[  1] +96*u[  2]-36*u[  3]+
          32/3*u[  4]-3/2*u[  5]-50*ux[1]*dx);
```

In computing the second derivative at the left boundary, uxx[1], the first derivative at the left boundary is included, that is, ux[1]. In this way, a Neumann BC is accommodated (ux[1] is included in the input argument ux).

For nl=1, only values of the dependent variable (and not the first derivative) are included in the weighted sum.

```
if(nl==1)
  uxx[1]=r12dxs*
         (45*u[  1]-154*u[  2]+214*u[  3]-
          156*u[  4] +61*u[  5] -10*u[  6]);
```

The dependent variable at points i=1,2,3,4,5,6 is used in the FD approximation for uxx[1] to remain within the x domain (fictitious points outside the x domain are not used).

Six points are used rather than five (as in the centered approximation for uxx[i]) since the FD applies at the left boundary and is not centered (around i). Six points provide a fourth order FD approximation which is the same order as the FDs at the interior points in x.

Similar considerations apply at the upper boundary value of x with nu=1,2.

Appendix A3: Function van1

A listing of function van1 follows.

```
van1=function(xl,xu,n,u,v){
#
# Function van1 computes the van Leer
# flux limiter approximation of a first
# derivative
#
# Declare arrays
  ux=rep(0,n);phi=rep(0,n);r=rep(0,n)
#
# Grid spacing
  dx=(xu-xl)/(n-1);
#
# Tolerance for limiter switching
  delta=1.0e-05;
#
# Positive v
  if(v >= 0){
    for(i in 3:(n-1)){
      if(abs(u[i]-u[i-1])<delta){
        phi[i]=0;
      }else{
        r[i]=(u[i+1]-u[i])/(u[i]-u[i-1])
        if(r[i]<0){
```

239

```
        phi[i]=0;
      }else{
        phi[i]=max(0,min(2*r[i],min(0.5*
              (1+r[i]),2)));
      }
    }
    if(abs(u[i-1]-u[i-2])<delta){
      phi[i-1]=0;
    }else{
      r[i-1]=(u[i]-u[i-1])/(u[i-1]-u[i-2]);
      if(r[i-1]<0){
        phi[i-1]=0;
      }else{
        phi[i-1]=max(0,min(2*r[i-1],min(0.5*
        (1+r[i-1]),2)));
      }
    }
    flux2=u[i  ]+(u[i  ]-u[i-1])*phi[i  ]/2;
    flux1=u[i-1]+(u[i-1]-u[i-2])*phi[i-1]/2;
    ux[i]=(flux2-flux1)/dx;
  }
    ux[1]=(-u[1]+u[2])/dx;
    ux[2]=(-u[1]+u[2])/dx;
    ux[n]=(u[n]-u[n-1])/dx;
  }
#
# Negative v
  if(v < 0){
    for(i in 2:(n-2)){
      if(abs(u[i]-u[i+1])<delta){
        phi[i]=0;
      }else{
        r[i]=(u[i-1]-u[i])/(u[i]-u[i+1]);
        if(r[i]<0){
```

```
          phi[i]=0;
        }else{
          phi[i]=max(0,min(2*r[i],min(0.5*
               (1.0+r[i]),2)));
        }
      }
      if(abs(u[i+1]-u[i+2])<delta){
        phi[i+1]=0;
      }else{
        r[i+1]=(u[i]-u[i+1])/(u[i+1]-u[i+2]);
        if(r[i+1]<0){
          phi[i+1]=0;
        }else{
          phi[i+1]=max(0,min(2*r[i+1],min(0.5*
          (1.0+r[i+1]),2)));
        }
      }
      flux2=u[i  ]+(u[i  ]-u[i+1])*phi[i  ]/2;
      flux1=u[i+1]+(u[i+1]-u[i+2])*phi[i+1]/2;
      ux[i]=-(flux2-flux1)/dx
    }
      ux[1]=(-u[1]+u[2])/dx;
      ux[n-1]=(-u[n-1]+u[n])/dx;
      ux[n]  =(-u[n-1]+u[n])/dx;
  }
#
# All points concluded (x=xl,...,x=xu)
  return(c(ux))
}
```

The use of van1 is illustrated in Chapter 2, including an explanation of the arguments in van1=function(x1,xu,n,u,v). This limiter is relatively complicated (and nonlinear). Details are given as reference [2] in Chapter 2 which includes a discussion of the theory of flux limiters.

Appendix A4: Functions super, smart

A listing of function super follows.

```
super=function(xl,xu,n,u,v){
#
# Function super computes the superbee
# flux limiter approximation of a first
# derivative
#
# Declare arrays
  ux=rep(0,n);phi=rep(0,n);r=rep(0,n);
#
# Grid spacing
  dx=(xu-xl)/(n-1);
#
# Tolerance for limiter switching
  delta=1.0e-05;
#
# Positive v
  if(v >= 0){
    for(i in 3:(n-1)){
     if(abs(u[i]-u[i-1])<delta){
       phi[i]=0;
      }else{
       r[i]=(u[i+1]-u[i])/(u[i]-u[i-1]);
       phi[i]=max(0,min(2*r[i],1),
```

243

```
                    min(r[i],2));
        }
        if(abs(u[i-1]-u[i-2])<delta){
          phi[i-1]=0;
        }else{
          r[i-1]=(u[i]-u[i-1])/(u[i-1]-u[i-2]);
          phi[i-1]=max(0,min(2*r[i-1],1),
                  min(r[i-1],2));
        }
        flux2=u[i  ]+(u[i  ]-u[i-1])*phi[i  ]/2;
        flux1=u[i-1]+(u[i-1]-u[i-2])*phi[i-1]/2;
        ux[i]=(flux2-flux1)/dx;
    }
    ux[1]=(-u[1]+u[2])/dx;
    ux[2]=(-u[1]+u[2])/dx;
    ux[n]=(u[n]-u[n-1])/dx;
  }
#
# Negative v
  if(v <0){
    for(i in 2:(n-2)){
      if(abs(u[i]-u[i+1])<delta){
        phi[i]=0;
      }else{
        r[i]=(u[i-1]-u[i])/(u[i]-u[i+1]);
        phi[i]=max(0,min(2*r[i],1),
                min(r[i],2));
      }
      if(abs(u[i+1]-u[i+2])<delta){
        phi[i+1]=0;
      }else{
        r[i+1]=(u[i]-u[i+1])/(u[i+1]-u[i+2]);
        phi[i+1]=max(0,min(2*r[i+1],1),
                  min(r[i+1],2));
```

```
      }
      flux2=u[i  ]+(u[i  ]-u[i+1])*phi[i  ]/2;
      flux1=u[i+1]+(u[i+1]-u[i+2])*phi[i+1]/2;
      ux[i]=-(flux2-flux1)/dx;
    }
    ux[1]=(-u[1]+u[2])/dx;
    ux[n-1]=(-u[n-1]+u[n])/dx;
    ux[n  ]=(-u[n-1]+u[n])/dx;
  }
#
# All points concluded (x=xl,...,x=xu)
  return(c(ux))
}
```

A listing of function super follows.

```
  smart=function(xl,xu,n,u,v){
#
# Function smart computes the smart
# flux limiter appproximation of a
# first derivative
#
# Declare arrays
  ux=rep(0,n);phi=rep(0,n);r=rep(0,n)
#
# Grid spacing
  dx=(xu-xl)/(n-1);
#
# Tolerance for limiter switching
  delta=1.0e-05;
#
# Positive v
  if(v >= 0){
    for(i in 3:(n-1)){
      if(abs(u[i]-u[i-1])<delta){
```

```
        phi[i]=0;
      }else{
        r[i]=(u[i+1]-u[i])/(u[i]-u[i-1]);
        phi[i]=max(0,min(4,0.75*r[i]+0.25,
                2*r[i]));
      }
      if(abs(u[i-1]-u[i-2])<delta){
        phi[i-1]=0;
      }else{
        r[i-1]=(u[i]-u[i-1])/(u[i-1]-u[i-2]);
        phi[i-1]=max(0,min(4,0.75*r[i-1]+0.25,
                2*r[i-1]));
      }
      flux2=u[i  ]+(u[i  ]-u[i-1])*phi[i  ]/2;
      flux1=u[i-1]+(u[i-1]-u[i-2])*phi[i-1]/2;
      ux[i]=(flux2-flux1)/dx;
    }
    ux[1]=(-u[1]+u[2])/dx;
    ux[2]=(-u[1]+u[2])/dx;
    ux[n]=(u[n]-u[n-1])/dx;
  }
#
# Negative v
  if(v < 0){
    for(i in 2:(n-2)){
      if(abs(u[i]-u[i+1])<delta){
        phi[i]=0;
      }else{
        r[i]=(u[i-1]-u[i])/(u[i]-u[i+1]);
        phi[i]=max(0,min(4,0.75*r[i]+0.25,
                2*r[i]));
      }
      if(abs(u[i+1]-u[i+2])<delta){
        phi[i+1]=0;
```

```
    }else{
      r[i+1]=(u[i]-u[i+1])/(u[i+1]-u[i+2]);
    phi[i+1]=max(0,min(4,0.75*r[i+1]+0.25,
              2*r[i+1]));
    }
    flux2=u[i  ]+(u[i  ]-u[i+1])*phi[i  ]/2;
    flux1=u[i+1]+(u[i+1]-u[i+2])*phi[i+1]/2;
    ux[i]=-(flux2-flux1)/dx;
  }
  ux[1]=(-u[1]+u[2])/dx;
  ux[n-1]=(-u[n-1]+u[n])/dx;
  ux[n  ]=(-u[n-1]+u[n])/dx;
  }
#
# All points concluded (x=xl,...,x=xu)
  return(c(ux))
}
```

super and smart have the same calling sequence as vanl in Appendix A3. Therefore, they can be used interchangeably with vanl by merely changing the name of the function where they are called.

Details for super, smart are given as reference [2] in Chapter 2 which includes a discussion of the theory of flux limiters.

Index

Printed in the United States
By Bookmasters